TECHNICAL COMMUNICATION QUARTERLY, 15(3), 285–292

W0018674

Guest Editor's Introduction: Communication in Technology Transfer and Diffusion: Defining the Field

Nancy W. Coppola
New Jersey Institute of Technology

This issue of *Technical Communication Quarterly* provides an introduction to our field's connections with technology transfer and diffusion. Technology transfer, the complex social process that moves technology from bench to market, drives global economic growth; technology diffusion, the market-driven process by which innovations are adopted and implemented, follows similar patterns. Indeed, technology transfer and diffusion may be considered synonymous with the phenomenon of growth in a global economy.

Such patterns of growth are pervasive. The National Science Foundation's biennial report, *Science and Engineering Indicators 2004* (National Science Board, 2004) showed that, although the United States continues to provide a singular benchmark for research and development (R&D), global R&D spending is advancing with more international research collaboration and alliances than ever before. Increasing government funding in defense, health, and counterterrorism, as well as more than 20 federal laws and directives, promotes technology transfer initiatives. In other market measures, patent issues rose overall with new academic patent applications quadrupling between 1991 and 2001. Universities that see technology transfer as a venue for marketing intellectual property are also home to growing multidisciplinary scholarship in technology transfer. A new journal published by The Johns Hopkins University Press for the Colorado Institute for Technology Transfer and Implementation is but one example. The editors of that journal, *Comparative Technology Transfer and Society* (Seely, Klein, & Klingner, 2003), cited growth in technology transfer scholarship as justification, in part, for the journal's inauguration, noting a 2002 database search that produced several thousand items under the heading *technology transfer* (Seely, 2003). Rogers (2003), the field's bibliographer of diffusion of innovations, noted the growth of diffusion studies in his classic work, *Diffusion of Innovations*, now in its fifth edition. In 1962, Rogers found only 405 publications in the diffusion field; in 2003, he estimated more than 5,000 works. "No other field of behavior science research" he reported, "represents more effort by more scholars in more disciplines in more nations" (Rogers, 2003, p. xviii).

HISTORY AND CHALLENGE

Although it may be argued that technology transfer and diffusion are embedded in the current phenomenon of growth in a global economy, technology transfer and technical communication have been intertwined since the time when *homo erectus* created tools and needed to talk about their use. Chinese campfire chats in Renzi Cave 2.25 million years ago, empire building, military conquests, invention, trade, and spying—all involve technology transfer and diffusion. Technology transfer historian Seely (2003) delineated the major modern historical instances of organized technology transfer: World War II, the Marshall Plan, the end of European colonialism, the Cold War, the dawning of the Space Age, and the various technological advances that marked the second half of the 20th century.

Not surprisingly, organized technology transfer and professional technical communication have similar ontological beginnings that follow the megascientific growth in the post-World War II era of Big Science. That enterprise's concomitant need for technology transfer and for those who could write about technology pushed our field from an activity to a profession (Coppola & Elliot, 2005). Following World War II, the first technological war that required technical communication, wartime technologies were translated into peacetime use; defense contractors responded to government requirements with a battery of proposals, reports, and procedures; and new technically based consumer products required manuals (Connors, 1982).

In the early 21st century, technology transfer and diffusion and technical communication are facing similar challenges from current political, social, and economic events. *Science and Economic Indicators 2004* (National Science Board, 2004) gave a differentiated picture of outsourcing, collaboration, and globalization. Outsourcing of R&D is a trend; the growth rate of contract R&D (or company-funded R&D performed externally) in recent years exceeded that of company-funded R&D performed in-house. As well, the complexities and costs of high-technology innovations encourage collaboration among domestic and international researchers; space activities, and the International Space Station in particular, are examples of a venture so expensive and complex that nations must pursue collaboration. Cutting-edge research that can lead to marketable innovation is most frequently published in the world's leading journals by scholars in China, South Korea, Malaysia, Singapore, and Taiwan. Major U.S. companies that produce high-technology products are regularly setting up units in and routing their intellectual property and financial fruits to Ireland and other low-tax countries

Technical Communication Quarterly

Volume 15, Number 3   Summer 2006

TCQ Reviewers

TCQ relies on the expertise of our reviewers not only to select articles for publication but also to help authors see possibilities for development. We thank them for their contributions to the quality of the journal.

(Simpson, 2005). Pursuit of rapid technology turnaround and lower costs draw manufacturing processes to Mexico and China.

Technical communication is also at the crossroads of offshoring and out-sourcing. Whereas in technology transfer the highly skilled work of technology innovation is taking up global residency (Brown, Green, & Lauder, 2001), the more routinized aspects of technical communication are moving offshore. *Technical Communication*'s (*TC*) August issue annually forecasts the state of our profession and its future. *TC*'s editor Hayhoe (2005) described our current challenges:

> In North America and increasingly in Europe, we have seen tasks that were formerly performed in our countries increasingly outsourced to the fledgling developing economies. Here in the U.S., textile and heavy manufacturing have virtually disappeared, and calls to tech support or customer service numbers are more often answered in Bangalore than in Boston. Not surprisingly, documentation is among the tasks that have been offshored. (p. 265)

ELEMENTALISM AND BEYOND

The problem goes deeper than offshoring goods and services. Consider the three traditional models of technology transfer found in the literature and summarized by Williams and Gibson (1990). First, the appropriability model, with its emphasis on technology development, follows the logic that good technologies sell themselves, and no deliberate transfer strategies are needed. Second, the dissemination model, with its emphasis on technology acceptance, features the expert informing potential users of the technology. Once linkages are formed, the new technology will flow from expert to nonexpert. Third, the knowledge utilization model, with its emphasis on technology application, begins to consider communication strategies between research and client as the basic research moves in one direction to become a developed idea and then product.

These techno-centric models reflect an elementalist view of technology and a simplistic approach to technology transfer that assumes a one-way, chronological process. Williams and Gibson (1990) were perhaps the first to note the limitations of these models and their source–destination paradigm as the editors proposed a communication model that is ongoing, iterative, and pervasively responsive. But it was Doheny-Farina (1992) who set the field for us with *Rhetoric, Innovation, Technology: Case Studies of Technical Communication in Technology Transfers*. Doheny-Farina was the first to unpack the technology transfer and diffusion literature, with its sophisticated understandings of manufacturing, R&D, marketing, and industry, to discover the underlying simplistic view of communication. Doheny-Farina argued for a rhetorical, socially constructed communication model with every aspect of technology transfers being negotiated, constructed, and reconstructed by the participants: "Those who will participate in or study technology transfers, the commercialization of innovations, the dissemination or diffusion of ideas from labs to markets should come to some understanding of the rhetorical processes that are ever-present in these enterprises" (p. 30).

Following the initial push for technology transfer provided by the Stevenson–Wydler Technology Innovation Act of 1980 and the ensuing attention to the enterprise by scholars, many studies began to include the importance of communication (Allen, 1984; Tushman & Nadler, 1986). Technology transfer suddenly became a relevant, interesting topic to professional communicators (Perkins, 1993, 1994; Roberts, 1991). Yet, after the mid-1990s, technical communication publications dwindled to a few isolated cases (Irwin, 1996), a handful of conference papers (Liebetrau, 1994), and passing references critical of the social constructivist approach used in technology transfer (Coogan, 2002; Longo, 1998). Williams and Gibson's (1990) book went out of print. Yet other fields have continued to recognize the essential role that communication plays in transferring technology. A literature review revealed just a sampling of the many disciplines that acknowledge communication's role in enabling technology transfer and diffusion—agriculture (Postlewait, Parker, & Zilberman, 1993), aerospace (Ornatowski, 1998), defense industry (Van Nostrand, 1997), electronic engineering (Allen, 1984), and software (Amsden & Amsden, 1993). Nevertheless, these approaches do not get to the heart of communication because they base their analyses on the concept of information transfer. The *Journal of Engineering and Technology Management* attempted to dispel an instrumentalist perspective in its special 2004 issue devoted to the human side of technological innovation, yet its article on communication in new product development still worked from an information-processing model (Patrashkova & McComb, 2004).

The metaphorical framework that dominates the literature of technology transfer is one of the barriers to overcome and boundaries to span. The inaugural issue of *Comparative Technology Transfer and Society* set its view of technology transfer as a movement across national, geographic, social, and cultural boundaries. A review of the literature on barriers to successful technology transfer by Greiner and Franza (2003) classified barriers into three main categories—technical, regulatory, and human. Not surprisingly, the human-based barriers (embodying and articulating technical and regulatory barriers) are built from communication problems. According to Williams and Gibson (1990), when people of widely varying backgrounds, such as those found in R&D consortia and in university–industry or government–industry partnerships, attempt to share information and technology with one another, there are communication barriers. Differences in vocabulary, ethos, and motivations amount to a communication effort between what are essentially different cultures (p. 11).

Establishing a barrier model, however, prevents us from fully appreciating the subtle complexities of technology transfer and diffusion. Linguistic and cultural differences are pervasive and persistent. They are never overcome so much as they are worked through (Killingsworth, 2005, p. 256). The idea of engaging subtlety rather than vaulting barriers suits this special issue. Because technology transfer and diffusion embody movement and flow, the metaphor of navigation might provide a more realistic conceptual framework. Technology transfer might best be understood as a navigating practice; like wily Odysseus, we are not mastering but negotiating the process. Borrowing from Francis Bacon's *ars tradendi* (Spedding, Ellis, & Heath, 1879/1996), we might also think about technology transfer as an art rather than a science. *Ars* suggests the creativity that Richard Florida (2002) described in his creative cohort of scientists, engineers, architects, educators, writers, artists, and entertainers who contribute to the new economy by creating new ideas, new technology, and new content. When Bacon coined the phrase, he intended it to exemplify this point—that what we know depends on the practices of communication by which the knowledge comes to us. The Victorian translation of Bacon's *ars tradendi* (Spedding et al., 1879/1996) as *the arts of transmission* resonates with modern scholars (Chandler, Davidson, & Johns, 2004) in examining different practices of knowledge making. But *tradendi* is also *surrender*, the art of surrender, and nicely exemplifies the metaphor of navigating and negotiating the sea of change involved in technological innovation. And so for us in technical communication, the art of transmission helps us rethink elementalist assumptions about technology transfer as more than commoditization of research. Beginning with the history of innovation in this special issue, Gregory Wickliff dispels the idea that Daguerre was the first to transfer phototechnology. Wickliff shows us that many innovative minds—pioneers in France, England, and the United States—were at the forefront of photography at the same time. The reification of innovation is carefully articulated by Katherine Durack, who explores relationships between the peer review and patent systems. Durack cautions those of us who see science publication as the only route to knowledge validation that we need to be equally familiar with the increasing patent activities in our universities. Imagine the complex problem solving involved when 30 technologists negotiate transfer and large-scale diffusion of a sophisticated innovation—Internet2. Barbara Mirel and Nicholas Johnson demonstrate that new models of diffusion research are needed that consider group social dynamics to achieve technical improvements in an innovation. David Dayton gives us a hybrid theoretical framework to guide innovation adoption in work groups. Focusing on a critical technology for our field—a single-source content-management system—Dayton demonstrates the framework's validity as he applies it to a case of a technical communication work group. Finally, the role of technical writing in international technology transfer is examined by Barry Thatcher in his case study of four U.S.-owned manufacturing plants in Mexico. His cross-cultural inquiry unmasks conflicting assumptions by plant owners and plant workers, revealing the prevalence of traditional Mexican cultural and rhetorical traditions.

CHARTING NEW DIRECTIONS

Communication in technology transfer and diffusion provides us with at least one solid answer to perennial questions about future directions for our field. As a new site of practice, technology transfer and diffusion represent opportunities for engagement outside of fields that we have already colonized, such as information systems. Disciplines other than technical communication are publishing models of innovation development that include communication and are doing a poor job of representing us. It is time for us to chart new directions in pedagogy, research, and practice.

Pedagogy

Clearly, the way to deepen our place in a market-oriented productive economy is to prepare better educated and enlightened technical communication professionals. As science and technology development becomes more interdisciplinary and as the number of U.S. citizens who are training to become scientists and engineers declines, technical communication courses are a natural home for motivation and learning about a new product-development process. Many of us who work in universities might develop working relationships with our technology transfer officers and research center directors; thus, we could provide opportunities for our students to be part of science in the making, as Latour and Woolgar (1986) suggested.

Research

Diffusion research methods, established 60 years ago, have become so widely recognized and applied that Rogers (2003) called on diffusion scholars to move beyond these proven methods to adopt a more critical stance. On the other hand, technology transfer scholars need to develop robust theories and methods. Scholarly study of technology transfer lacks a widely accepted theoretical conceptualization and an effective methodology. Most of the technology transfer literature consists of case studies and deals with pragmatic questions not associated with theory. Rather than focusing on barriers and failures, we need models in which the dependent variable is successful technology transfer and the independent variables encompass the subtleties of language.

Practice

Technical communicators, who are masters of contingent flexibility, are poised to excel in these fields of rapid technological change, globalization of business, and fast-moving markets. The high failure rate among international technology transfer (Munir, 2002) is alarming to many organizations; these groups could certainly benefit from a communication strategist with experience in the global information economy. As Giammona (2004) found in her survey of technical communicators, we bring unique gifts to our role in innovation: "The gift of understanding how pieces fit into the whole is how we add value to the innovative process" (p. 355).

I hope that you will join me and the authors of the articles in this special issue as we chart new directions, maintain an open flow of ideas in our global environment, and embrace innovation.

ACKNOWLEDGMENTS

I am grateful for the insight of this special issue's peer reviewers: Joseph Chew, Lawrence Berkeley National Laboratory; Stephen Doheny-Farina, Clarkson University; and Norbert Elliot, New Jersey Institute of Technology. I thank graduate student Amy Malsbury Nowak for her editing assistance.

REFERENCES

Allen, T. J. (1984). *Managing the flow of technology.* Cambridge, MA: MIT Press.

Amsden, D. C., & Amsden, A. A. (1993). The KIVA story: A paradigm of technology transfer. *IEEE Transactions on Professional Communication, 36,* 190–195.

Brown, P., Green, A., & Lauder, H. (2001). *High skills: Globalization, competitiveness, and skill formation* (1st ed.). New York: Oxford University Press.

Chandler, J., Davidson, A. I., & Johns, A. (2004). Arts of transmission: An introduction. *Critical Inquiry, 31,* 1–6.

Connors, R. J. (1982). The rise of technical writing instruction in America. *Journal of Technical Writing and Communication, 12,* 329–352.

Coogan, D. (2002). Public rhetoric and public safety at the Chicago Transit Authority: Three approaches to accident analysis. *Journal of Business and Technical Communication, 16,* 277–305.

Coppola, N. W., & Elliot, N. (2005). Big science or bricolage: An alternative model for research in technical communication. *IEEE Transactions on Professional Communication, 48,* 261–268.

Doheny-Farina, S. (1992). *Rhetoric, innovation, technology: Case studies of technical communication in technology transfers.* Cambridge, MA: MIT Press.

Florida, R. (2002). *The rise of the creative class: And how it's transforming work, leisure, community and everyday life.* New York: Basic Books.

Giammona, B. (2004). The future of technical communication: How innovation, technology, information management, and other forces are shaping the future of the profession. *Technical Communication, 51,* 349–366.

Greiner, M. A., & Franza, R. M. (2003). Barriers and bridges for successful environmental technology transfer. *Journal of Technology Transfer, 28,* 167–177.

Hayhoe, G. F. (2005). The future of technical communication. *Technical Communication, 52,* 265–266.

Irwin, H. (1996). The role of personal communication in technology transfer in the aerospace industry. *Australian Journal of Communication, 23*(2), 34–50.

Killingsworth, J. M. (2005). Rhetorical appeals: A revision. *Rhetoric Review, 24,* 249–263.

Latour, B., & Woolgar, S. (1986). *Laboratory life: The construction of scientific facts.* Princeton, NJ: Princeton University Press.

Liebetrau, S. F. (1994). The exhiliarating, dangerous world of technology transfer communication: Success and sanity through teamwork. *IPCC '94 Proceedings,* 23–27.

Longo, B. (1998). An approach for applying cultural study theory to technical writing research. *Technical Communication Quarterly, 7,* 53–73.

Munir, K. A. (2002). Being different: How normative and cognitive aspects of institutional environments influence technology transfer. *Human Relations, 55,* 1403–1428.

National Science Board. (2004). *Science and engineering indicators 2004* (Publication No. NSB-04-1). Arlington, VA: National Science Foundation.

Ornatowski, C. M. (1998). 2 + 2 = 5 if 2 is large enough: Rhetorical spaces of technology development in aerospace engine testing. *Journal of Business and Technical Communication, 12,* 315–342.

Patrashkova, R. A., & McComb, S. A. (2004). Exploring why more communication is not better: Insights from a computational model of cross-functional teams. *Journal of Engineering and Technology Management, 21,* 83–114.

Perkins, J. M. (1993). Social perspectives on technology transfer. *IEEE Transactions on Professional Communication, 36,* 185–189.

Perkins, J. M. (1994). Reconsidering technology transfer: An antifoundational perspective. In B. Sims (Ed.), *Studies in technical communication: Selected papers from the 1993 CCCC and NCTE meetings* (pp. 81–88). Denton, Texas: University of North Texas.

Postlewait, A., Parker, D. D., & Zilberman, D. (1993). The advent of biotechnology and technology transfer in agriculture. *Technological Forecasting and Social Change, 43,* 271–287.

Roberts, S. (1991). Technology transfer: An opportunity for technical communicators. *Technical Communication, 3,* 336–344.

Rogers, E. M. (2003). *Diffusion of innovations* (5th ed.). New York: Free Press.

Seely, B. E. (2003). Historical patterns in the scholarship of technology transfer. *Comparative Technology Transfer and Society, 1,* 7–48.

Seely, B. E., Klein, G., & Klingner, D. E. (2003). A note of welcome from the editors. *Comparative Tehcnology Transfer and Society, 1,* 1–2.

Simpson, G. R. (2005, November 7). Irish unit lets Microsoft cut taxes in U.S., Europe. *Wall Street Journal,* p. 1.

Spedding, J., Ellis, R., & Heath, D. (Eds.). (1996). *Collected works of Francis Bacon.* London: Routlege/Thoemmes. (Original work published 1879)

Tushman, M., & Nadler, D. (1986). Organizing for innovation. *California Management Review, 28*(3), 74–92.

Van Nostrand, A. J. (1997). *Fundable knowledge: The marketing of defense technology.* Mahwah, NJ: Lawrence Erlbaum Associates, Inc.

Williams, F., & Gibson, D. V. (Eds.). (1990). *Technology transfer: A communication perspective.* Newbury Park, CA: Sage.

Nancy W. Coppola is an associate professor of English at New Jersey Institute of Technology, where she directs the graduate program in professional and technical communication. Her edited volume, with Bill Karis, *Technical Communication, Deliberative Rhetoric, and Environmental Discourse: Connections and Directions,* is part of the ATTW Contemporary Studies in Technical Communication. Coppola has published numerous articles and book chapters on technical communication research, pedagogy, and assessment.

Light Writing: Technology Transfer and Photography to 1845

Gregory A. Wickliff
University of North Carolina at Charlotte

This article reviews the history of photography to 1845 in France, England, and the United States, emphasizing roles of collaboration, legal protection, and training in the development and transfer of the technologies of the heliograph, physautotype, daguerreotype, and calotype. It argues that early innovative work in photography was motivated by plural desires: to photo-illustrate printed publications, to capture scenes from nature, to render human portraiture, and to investigate scientific theories of radiation.

Amid narratives of the invention of photography, historians recognize at once the 1839 claims to invention of Louis Jacques Mandé Daguerre in France, and the rival claims of William Henry Fox Talbot in England. But other of the earliest work toward successful photography remains less well recognized and understood. In photography, many of the earliest technological innovations were fostered by collaboration, by legal protection, and by training. Amid abundant national and even nationalistic histories of photography, the international development and dissemination of early photo technology remains somewhat obscure. Although the images of Daguerre appeared magical or miraculous to most in 1839, they in fact represented the culmination of more than 20 years of incremental work toward a satisfactory process of photography. Permanent silver-based photo images evolved from a search for a photolithographic process. With improvements that shortened exposure times from days to minutes, photo technology evolved by 1839 to the point where it could render recognizable miniature images directly from nature without conflating the movements of shadows cast by the sun. In the months immediately before and following the publication of Daguerre's (1839/1971) how-to manual *Historique et Description des Procédés du Daguerréo Type et du Diorama* in August 1839, technical innovation and legal restrictions became prominent, working both as incentives and barriers to international transfer of the technology. As exposure times were shortened further, successful human portraiture and a host of early scientific and artistic applications were realized. The resulting early images represent a realization of some of the strongest latent desires in Western culture of this period: to produce photolithographs, to capture minutely detailed scenes from nature, to render economical and rapid human portraiture, and to scientifically study the nature of radiation itself.

What Marien (1997) termed the *prehistory* of photography, the period before 1839, may never be completely clear. The number of early 19th-century photo experimenters is large, the historical archives are incomplete, and as Foresta (1996) pointed out, until quite recently, most photo historians have been more interested in photography's second century than in its first. What can be done is to paraphrase and to compare early theories of image making that guided the practices of pioneers, the roles of collaboration, contractual negotiation, legal protection, and training in the early development and use of photography—especially in France, England, and the United States. Research is also needed on the early history of innovation and technology transfer in the rest of Europe as well as in South America, Central America, India, Australia, and elsewhere around the globe.

This article summarizes the major technological and rhetorical approaches to innovation of some of the earliest and best known photographic pioneers and acknowledges the contributions of some lesser known collaborators. Included are discussions of the work in France of the photographic pioneers Joseph-Nicephore Niépce and Daguerre, of the opticians Charles and Vincent Chevalier, and of the Parisian etcher Augustin Lemaitre. In England, Niépce first appeared with Francis Bauer, a botanist and fellow of the Royal Society. Paper-based photography moved forward in the 1830s with William Henry Fox Talbot, Sir John Herschel, and Nicolas Henneman. In the United States, Robert Cornelius, Joseph Saxton, Alexander Wolcott, John Johnson, William Henry Draper, and, years before his most famous long-distance telegraph demonstration, Samuel F. B. Morse all conducted simultaneously innovative work. Internationally, many of these same figures crossed borders in the first years of photography, carrying with them expertise, equipment, contracts, patent applications, and training materials.

A ROMANTIC AND TRADITIONAL HISTORY

The invention and dissemination of photography has most widely been written of as a triumph of an individual genius: Daguerre. This triumph is most closely associated with a specific date, August 19, 1839, when François Arago, a French astronomer and statesman, announced the details of the photographic process before an open meeting of the French Academy of Sciences and the Academy of Fine Arts. That June, Arago had already successfully argued the case before the French government for a lifetime pension for Daguerre in return for a renouncement of international patent rights. From a modern viewpoint, the culminating event that August is often narrated as a triumph of aesthetics and of individual ingenuity: the successful realization of the urge to make images with ever more verity. In late 1839, specimens of the new photographic process were publicly displayed, and the process was demonstrated, published, patented, and licensed. Devices were manufactured and advertised, training presented, and issues of scientific priority were raised and largely resolved. The August 19 event was publicized, attended by hundreds, and reported in all the major newspapers of the world. At the mere hint of the possibility of human portraiture through this new technology, an enormous implicit demand was at once created.

But earlier in the winter of 1839, Daguerre's scientific priority had already been challenged by Talbot in England, who, since 1834, had been experimenting with his own paper-based photographic processes (Schaaf, 1992, p.3). Written evidence of Daguerre's contractual work with Niépce proved that their collaborative experiments dated back to 1829, and the superior quality of Daguerre's photo images quickly prevailed with both scientific and popular audiences. For the first 20 years of photography, before the development of a successful wet-plate glass-negative process, *daguerreotype* was largely synonymous with photography itself. According to the dominant historical narrative, the process of photography was discovered by Daguerre, and the August 1839 presentation of the process to the French government completed the contractual purchase of the rights to the invention. France then made a gift of the process to the world—a gift that Daguerre's champion in the scientific community, Arago, promised would be both artistic and scientific.

REVISIONIST HISTORIES

What this thumbnail version of the traditional historical narrative leaves out is much indeed: much about the history of the supporting work in science and technology that led to photography; much about Daguerre's trials, collaborations, and contractual negotiations with Niépce, who had been at work on an effective process for photolithography since 1816; and much about how patent and license restrictions functioned to limit both early daguerreotype in England and early photography on paper in France. At the same time, the absence of similar patent restrictions in the United States allowed photographic innovation and photo commerce to flourish during the era of *daguerrian* popularity.

It appears that Daguerre, as an act of individual invention, did develop the final details of his imaging process independently between the years 1833 and 1837. But he did so by building on years of formally contracted collaboration with Niépce, who, with his brother Claude, had been at work on the problems of photolithography and photography for 10 years before he corresponded with and met Daguerre. Niépce's unexpected death from a stroke in 1833 allowed Daguerre, who was still contractually bound to Niépce's son Isidore, to continue experiments independently and to attach his own name to the improved version of photography that he presented to the world in 1839. French and British political rivalry in this historical period led to restrictive licensing practices for daguerreotype in Great Britain, to restrictions for photography on paper in France, and to debates before and between the British Royal Society and the French Academy of Sciences. Those in the United States, free from international patent and licensing restrictions on daguerreotypy, moved quickly to adopt and improve upon the technology, and they soon took the lead in several areas of imaging technology, shortening exposure times and making a succession of early photo-imaging firsts, especially in portraiture. Almost immediately, the technological innovations were traveling internationally: French daguerreotype training was conducted in England and America; French cameras and lenses were exported; English entrepreneurs worked with Talbot's paper process and trained practitioners in France; American innovative camera designs and plate sensitizing processes were exported to England.

With recent historiography, attempts have been made to rectify what Tranchtenberg termed in 1980, "a neglect of the intellectual history of the medium, a history of ideas about photography" (p. vii). Since then, critics have responded with Marxist histories of photo history that emphasize the power exerted by photographers over their often vulnerable human subjects, and these histories attend to the division of labor in commercial image making. Feminist histories have stressed the importance, aesthetics, and collaborations of women in photography, and have critiqued the exploitation of female images. Anthropological historians have revised earlier notions of photo-based ethnology and ethnography. Postmodern new historians have critiqued Romantic theories of light and agency, diminishing claims for any photo-based naturalistic truth or realism. Instead, they have emphasized the nationalistic, economic, and self-interests of photo pioneers and other photo historians. Despite this important work, the earliest history of photography, the period before 1839 and the first few years thereafter, remains fragmentary in the outline of its driving ideas and the social structures that fostered technological innovation.

Despite the precise dates of Arago's announcements on behalf of Daguerre, it has consistently proven difficult to know just where to begin a history of photography. Some historians have looked to the Greeks and to the notion of the camera obscura, the dark room with its small aperture that projects an exterior scene on an interior wall. Aristotle's writing *Problemata* from about 350 B.C. contains passages about using a camera-like device for observing eclipses. Other histories begin in the 18th century with photochemistry and the work of Johann Heinrich Schulze in Germany, and the work of Sir Humphry Davy and Thomas Wedgwood in England. Each had experimented with the darkening of silver compounds in the presence of light, but none had been able to arrest or fix the resulting shadowy images. Readers may also consider a host of both documentary and apocryphal stories of the first photographic successes: the claims for priority of Hippolyte Bayard in France or the earlier work of Elizabeth Fulhame in England who in 1794 described a silver-based process for duplicating maps that Schaaf (1992) called *photographic*. As Batchen (1997, p. 35) pointed out, modern historians of photography have perhaps too often imitated their subjects by championing individuals as individual inventors of photography or photographic concepts. What is clear is that there was a widely felt urge to create images photochemically within industrializing nations in the early 19th century.

TECHNOLOGY AND THE URGE TO IMAGE

By the early 19th century, improved imaging technologies were in great demand across Europe. As Rosenblum (1997) wrote, "By the time it was announced in 1839, Western industrialized society was ready for photography" (p. 7). Schaaf (1992), writing of the work in photochemistry of Wedgwood and Davy, was more explicit. He argued that late 18th- and early 19th-century work in photochemistry is striking because it reveals

> an insistent desire to photograph despite an inability to actually bring it to fruition. It suggests that desiring coincided with scientific discovery rather than flowed directly from it. Photography's origins lie therefore not in a single causal chain or moment of individual genius but in a complex knot of historical and cultural forces that can be glimpsed from afar but may never be fully unraveled. (pp. 24–25)

Wakeman (1973), a print historian, argued that the urge to mechanize and improve imaging was broadly felt in Western cultures and that it lies at the center of the modernist impulse itself. As printing multiplied text, the desire to multiply and improve illustration increased in proportion. Wakeman wrote that "all the important progress made in the printing of pictures during the Victorian period shows constant attempts to reduce the importance of the artist as an intermediary, and at the same time to reproduce the tones seen in nature" (p. 12).

Sontag (1977) went even further, arguing that the photographic impulse was and remains, in fact, fundamentally not only modern, but also democratic, no matter what the mundane subject. For Sontag, as for Wakeman, the increased number and range of photo images, beyond those made through painting and engraving, signified an essential characteristic of the popular technology from 1839 forward:

> From its start, photography implied the capture of the largest possible number of subjects. Painting never had so imperial a scope. The subsequent industrialization of camera technology only carried out a promise inherent in photography from its very beginning: to democratize all experiences by translating them into images. (p. 7)

The urge to image grew alongside popular literacy in the early 19th century. Visual artists had used the camera lucida, a rotating prism, to project images for sketching since 1807, and in the 1820s and 1830s, the portable camera obscura was widely used as a drawing aid. In 1798, Alois Solenfeld patented his process for lithography on fine-grained limestone, representing an early revolution in textual illustration. Engravings on wood had been supplemented by engravings on copper and on durable steel by 1830. The invention of the steam press by Frederick Koenig in 1814 dramatically increased the availability of texts and lowered production costs. Gains in basic print literacy were exponential, as the number and size of print periodical circulations increased rapidly in the first half of the 19th century. As early as 1816, these trends, especially in lithography, had drawn the attention of the two inventive French brothers, Claude and Joseph Niépce. Though successful designers and technicians in several fields, the two never became successful entrepreneurs.

For Joseph Niépce, perhaps the most diligent of the early photo pioneers, the initial goal appears to have been a process of photolithography for textual illustration. This may seem odd to our contemporary sensibility, raised and fed on naturalistic photographs, especially portraits, by the thousands, but improvements in lithography for document publication were more central to Niépce's earliest goals. From the very first, he also experimented with photo imaging on a wide range of media: paper, glass, stone, pewter, silver, and silver-plated bronze. His successes were many, including a positive–negative iodine-on-silver plate process that he developed with Daguerre, capturing images from nature that he termed *points-of-view*; exposure times at first were in days and later in hours. Daguerre's most fundamental contribution to the early work was to discover that the latent image left by a much shorter exposure time, of 3 to 30 min, could be brought out on silvered plates by mercury fumes.

Daguerre initially approached photography with goals that in some ways differed from those of Niépce. Daguerre was a visual artist and a stage designer, trained to paint operatic sets and huge popular panoramas. As an artist, his strongest interest was in lighting effects. By 1822, he had developed with Charles Marie Bouton an innovative theater that combined translucent paintings on gauze with lighting and sound effects set in motion in a theater without living actors (Buerger, 1989, p. 15). This was Daguerre's *diorama*, a kind of precursor to cinema consisting of a darkened room where the audience sat and viewed an illuminated moving image that appeared and dissolved on a painted translucent screen. Within the diorama, the image appeared to change as the lighting shifted front to back, simulating, for example, the transition from day to night. Buerger (1989) characterized Daguerre's prephotographic artwork as that of "a romantic painter with a severely naturalistic bent" (p. 13). For Daguerre, the initial promise of photography was an improved process of rendering miniature images as models for his theatrical presentations. In his earliest imaging experiments, it appears that Daguerre used light-sensitive phosphorescent materials, and the needle-etching process of *cliché-verre*: drawing on smoked glass.

For Niépce and for Daguerre, the initial urge to photograph represented these two responses to the demands for improved imaging: lithographic printing and cinematic public entertainment. Both were nascent cultural demands that would only grow throughout the 19th century. Once a satisfactory way to produce photo images was achieved, along with the apparatus, methods for plate preparation, exposure processes, developing processes, fixing processes, and shortened exposure times, commercial photography was quickly developed and widely disseminated. National and international distribution networks were established using the technology of steam presses for journalism and training materials, and steamships were used for the transportation of exports, information, and materials. Almost from the beginning, steam-powered machines were adapted to manufacture and polish the silver plates for photographic exposures. No mere timeline can adequately represent the web of relationships that imagined, fostered, developed, and disseminated the earliest photography.

THE HELIOGRAPH, PHYSAUTOTYPE, DAGUERREOTYPE IN FRANCE—FROM PHOTOLITHOGRAPHY TO SCENES FROM NATURE

In 2003, the Harry Ransom Research Library hosted an international conference, "At First Light—Reflections on the Invention of Photography." At its center resided what is at present considered the earliest surviving photograph, "View from the Window at Le Gras," made by Niépce in 1826 or 1827. It is one of five images left with the botanist Franz Bauer in England after a failed 1827 attempt by Niépce to get the British Court and a group of fellows of the Royal Society to recognize and financially support his early photographic work. Four of the images demonstrate Niépce's work in photolithography—the process he termed *heliography*. Niépce used heliography to duplicate engravings and paintings. The best known examples are Niépce's 1822 reproductions of an engraving of Cardinal Georges d'Amboise (see Figure 1). Less well known are Niépce's reproductions of other printed artwork, such as his "A Greek Man and Woman," one of five unreproduced images added to the collections of the Musée Niépce in July 2000. Most interesting, the latter image, said by Isidore Niépce to have been made before the association with Daguerre, is on a bronze, silver-coated plate, just as those that would later be made by daguerreotypy. Despite these successes, by 1828 Niépce wrote that he moved away from the reproduction of engravings to put more time into making satisfactory views from nature.

This image, taken from nature with the camera obscura (see Figure 2), now is celebrated as the first photograph; it was one of what Niépce termed a point-

FIGURE 1　An 1827 photo etching of an engraving of Cardinal Georges D'Amboise reproduced through heliography by Joseph Nicéphore Niépce. Prints such as this one first demonstrated the practicality of photolithography. From the Gernsheim Collection, Harry Ransom
Humanities Research Center, The University of Texas at Austin. Reproduced with permission.

FIGURE 2　"View from the Window at Le Gras." A point-of-view by Joseph Nicéphore
Niépce, ca. 1826, considered to be the oldest surviving photographic image from nature. From
the Gernsheim Collection, Harry Ransom Humanities Research Center, The University of
Texas at Austin. Reproduced with permission.

of-view. The image is a unique positive–negative on a polished pewter plate that presents a view of the outbuildings, courtyard, trees, and landscape as seen from an upstairs window at the Niépce estate. The image is recognizably a rooftop, but the shadows are confused by an extremely long exposure time of approximately 60 to 100 hr (Marignier, 1990, p.115). The image is now on permanent display at the Harry Ransom Research Center in a nitrogen-gas-filled and climate-controlled kiosk, a veritable icon of modernity. Compared to images with shorter exposure times, it could as easily be considered an imaging failure rather than an initial success (Maison Nicéphore Niépce, n.d.).

If Niépce made the first successful photo images from nature, his points-of-view, why are historians not writing about the Niépceotype process? The answer reveals a good deal about the history of the technologies in general and about the early 19th century legal and social systems in which photography first appeared. The short answer is that Niépce was not the successful entrepreneur that Daguerre was. He did not name his early imaging work after himself, but rather, he aimed for nomenclature that described the ideals he sought—*heliography* (sun-writing) for his lithographic processes and *physautotypy* (nature writing herself) for the work that we now think of as photographic. His description of the invention is clear in his *Memoire on the Heliograph*:

> The invention which I made and to which I gave the name "heliography" consists in the automatic reproduction, by the action of light, with their gradations of tones from black to white, of the images obtained in the camera obscura. (as cited in Trachtenberg, 1980, p. 5)

The story of heliography has a significant prelude of its own. Jean-Louis Marignier (1999; Bonnet & Marignier, 2003) has begun to document the entire range of Niépce's work. In 1807, Niépce and his brother Claude applied for a patent for an internal combustion engine—the first of its kind. The resulting boat motor, which they termed the *Pyreolophore*, was first powered by explosive lycopodium spores and later by resin-coated coal dust and fuel-injected kerosene. At the same time, the brothers developed a hydrostatic pump. In 1811, one of the effects of Napoleon's attempts to isolate England from wider trade was a shortage of indigo, the blue dye. The Niépce brothers recognized an opportunity in the national need and went to work in response to a government competition to extract blue dye from the native French woad plant. In 1818, Niépce created a prototype bicycle—he called it a *velocipede*—that had neither pedals nor chain-driven gears. In short, the brothers were serious inventors. In 1816, Claude traveled to England to seek backing for the development of the Pyreolophore engine. While Claude was away in England, Niépce turned his attention to improvements in lithography.

In 1798, Alois Senefelder discovered a successful process for engraving upon stone or lithography, and this discovery helped to fuel a demand for quality printed illustrations. In 1816, the French Company for the Encouragement for the National Industry launched a competition to find calcareous stones for lithography. Joseph Niépce responded to the call. In short, after their initial experiments in transportation and civil engineering, the Niépce brothers experimented with dye making and lithography—useful training for imaging pioneers.

By 1825, Joseph Niépce made successful heliographs in a process that used a bitumen (asphalt)-based substance. Working on glass, pewter, and stone, he succeeded in acid-etching photo-based plates, and with the help of printers like Augustin Lemaitre, he was then able to *pull* paper prints for duplication.

However, it was Joseph Niépce's collaboration with Daguerre that would ultimately prove most productive. Daguerre and Niépce had in common their relationships with the Parisian opticians Vincent and Charles Chevalier, to whom they both turned for lenses for their camera obscuras to fix the images of the camera. According to B. Newhall (1961), Charles Chevalier suggested that Daguerre write to Niépce in 1826. At first, surprised and worried that word of his experiments had become public, Niépce was not eager to collaborate. But, on his way to visit his brother Claude in England in December 1827, Niépce met with Daguerre and was impressed by the diorama.

Once in England, Niépce found that his brother Claude was dying, and with him, hope for any profit from their work on the internal combustion engine was dying. Niépce turned his attention to his successful imaging work and sought financial backing and scientific recognition for his accomplishments. In writing of Niépce's unsuccessful attempts to generate support in England during 1827 and 1828, Schaaf (1992) argued that although he sought out the right audiences for his invention, "Niépce's major failing seems to have been exceptionally unlucky timing" (p. 32). Officers of the British Royal Society who should have been prepared to recognize the value of his work in 1827 were ill, and some, including Niépce's brother, were dying. Consequently, Niépce appears to have left England empty-handed, leaving his images and his rhetorical case in the hands of Franz Bauer. Niépce and Daguerre met again upon Niépce's return in February 1828. Through correspondence, the two traded ideas. In 1829, Niépce asked Daguerre to enter into a partnership and to continue to develop heliography. They signed a contractual agreement in December 1829.

Daguerre soon brought a new material to the partnership—a distillate of lavender oil. By 1832, the collaborators developed a process for taking images from nature that they termed *physautotypes*. Developed over white petroleum, the resulting images could be viewed as either a negative or a positive, depending upon the angle of view, just as would the later daguerreotype process. The required exposure time was shortened to 7 or 8 hours in the sunlight.

The unexpected death of Niépce from a stroke in 1833 changed photo history deeply. Although in the initial work Niépce had taken the lead, Daguerre moved to the fore. Daguerre allowed Niépce's son Isidore to continue the collaboration as the contract specified, but the younger Niépce does not seem to have been an effective innovator or entrepreneur. Daguerre continued his own experiments until 1837, when he gained some confidence in the distinctive mercury-fumed process to which he gave his own name. At this point, he was without the undue fear of being considered in breach of his contract with the now deceased Niépce. The daguerreotype, with its richer detail and much shorter exposure time, was a distinct improvement upon Niépces' earlier heliography and their collaborative physautotype processes. Daguerre emphasized exposure time as a fundamental difference between his work and that of Niépce, writing

> I realized that the only way to succeed completely was to arrive at such rapidity that the impression could be produced in a few minutes, so that the shadows in nature should not have time to alter their position; and also that the manipulation should be simpler. (Trachtenberg, 1980, p. 12)

In January 1839, after a short, failed attempt to sell either individual or corporate subscriptions for the commercial purchase of the process, Daguerre and Arago, with the consent of Isidore Niépce, began to make their rhetorical application for a pension from the French government. Daguerre carefully guarded the secret of his process, fearing that once it was made public, no patent protection would be possible. The contract for Daguerre's pension was signed by the minister of the interior on June 14, 1839. The original contract specified equal shares for Niépce, Daguerre, and their heirs. In the final version, Daguerre received a larger pension than Isidore Niépce. Each received a pension of 4,000 francs for the work leading to the daguerreotype, but Daguerre was granted an additional 2,000 francs annually for disclosing the details of his work with the diorama, bringing his pension to the mandated ceiling of 6,000 francs. Because Joseph Niépce's work seemed to have been minimized by Arago in his appeal before the Chamber of Deputies, Isidore later felt that his father's pioneering work had been undercut.

Led by the recent work of Bonnet and Marignier (2003), scholarship focused on Niépce is expanding. Niépce's earliest heliographic processes and the collaborative physautotype processes have been successfully recreated. Images that were thought lost have been moved into Niépce Collections (Maison Nicéphore Niépce, n.d.). Yet controversy continues over the relative importance of the contributions made by Niépce and Daguerre to the invention of photography. Whatever the outcome, we know that Niépce was a serious inventor and innovator whose collaborative work in photography was first inspired by a desire for a photolithographic process. He first took up the work in response to government incentives, yet, despite the suggested relative security, the Niépce brothers consistently failed to find success in marketing their inventions. The development process proceeded very slowly: for the Pyreolophore engine (20 years) and for heliography and physautotypy (over 16 years before Joseph Niépce's death). Niépce's projects were developed collaboratively: first with his brother Claude, then with Lemaitre, and finally with Daguerre.

Daguerre's interest had begun in developing imaging effects. With Arago as a government advocate, he was also somewhat nationalistic. Daguerre patented his process in Britain through his agent, Miles Berry, one week before the process was publicly disclosed as "free to the world." Berry facilitated the sale of the patent rights for daguerreotype in England to Richard Beard, who then subsequently exercised a virtual monopoly in England under British Patent No. 8194 that did not expire until August 14, 1853.

Once the daguerreotype process was in the public domain, photographic experimentation and innovation spread extremely quickly. Harmant (1977) argued that "within thirty days of the meeting of the Academies of Sciences and the Fine Arts on 19 August 1839, Daguerre's process was known throughout Europe, thanks to the Paris newspapers and their circulation abroad" (p. 79). Daguerre was very quick to anticipate and to respond to the new demand created by news of the invention. In June 1839, he sold endorsements for subsidiary work such as the manufacture of cameras, the sales of training materials, and the endorsement of trainers:

> Publication of the Manual was actually covered by the terms of the original government agreement signed between Daguerre and Isidore Niépce. They then signed a further contract with Alphonse Girox, a curio and stamp dealer in the Rue de Coq-St. Honoré, granting him the right to sell the materials and equipment for the preparation of daguerreotype images. This contract is dated 22 June 1839. (Harmant, 1977, p. 79)

Together, Daguerre and Arago defended the daguerreotype against Talbot's claims of scientific priority. Like Talbot's images, Daguerre's earliest images were still lifes, landscapes, and architectural views—subjects that could be rendered well with long exposure times. Unlike Talbot's paper process, the daguerreotype presented images with amazing detail. A magnifying glass was often applied to plates to reveal additional levels of detail. Daguerre and Arago realized that future applications of the technology would reshape how people came to know the world in all its variety. Daguerre was not among the first to claim success with portraiture, despite the fascination with this use in the popular press from the first. However, Daguerre always recognized the wide range of important scientific applications and, perhaps even more important, the cultural applications:

> Everyone, with the aid of the DAGUERREOTYPE, will make a view of his castle or country-house; people will form collections of all kinds, which will be the more precious because art cannot imitate their accuracy and perfection of detail; besides, they are unalterable by light. Even portraits will be made, though the unsteadiness of the model presents, it is true, some difficulties [which need to be overcome] in order to succeed completely. (Trachtenberg 1980, p. 12)

In addition to Niépce and Daguerre, Hippolyte Bayard claimed in the spring of 1839 that he had independently developed a process of producing photos on paper. According to Schaaf (1992)

> Bayard had been working on a photographic process before 1839 and was able to show an independently invented paper negative before Talbot's working details were disclosed. Seeing the negative/positive approach as a disadvantage compared to Daguerre's, Bayard invented a direct positive process on paper by early spring 1839. Arago, however, persuaded him to keep the development to himself by supplying Bayard with photographic apparatus and encouraging him to develop it further. Daguerre's announcement then totally eclipsed Bayard's effort. (p. 177)

Keeler (1990) showed that Bayard and Talbot corresponded after 1839, and she suggested that Bayard may have attended the workshops given by Nicolaas Henneman, Talbot's assistant and collaborator in 1843 in France.

THE CALOTYPE IN ENGLAND
AND FRANCE—NATURE ON PAPER

For Talbot, who developed another version of photography completely independent of Niépce and Daguerre, photography first became a question of scientific priority. Second, it became a means to produce book art, and finally, it became a financially unsuccessful business venture, both internationally and at home. Talbot's initial process used sensitized paper to produce what he termed *photogenic drawings*, or what we now term *photograms*. He quickly moved forward with the help of Sir John Frederic William Herschel to develop and patent the first positive–negative paper-based photographic process: the calotype process. The negatives were the calotypes, from the Greek *kalos*, meaning beautiful, and the positives were simply termed *salted paper prints*. Like heliography, the calotype process afforded multiple prints from a single original, but because it used a translucent negative, it is a more direct precursor to modern film-based processes than daguerreotype with its unique images on metal plates. As a member of the English gentry, Talbot understood and protected his legal rights to this paper-based process in England through an 1841 patent. But it seems that he was not confident in conducting business in France, where he was willing to sell his rights to the calotype process without much financial gain. Schaaf (1992) argued that Talbot's motives for securing a patent were primarily to protect his claims for invention: "In later years, not so much for profit as in an effort to preserve his identity as the inventor, Talbot would patent subsequent photographic processes. This would bring him little fame, even less income, and no end of grief" (p. 42).

Talbot had experimented intermittently with photography on paper using silver salts between 1833 and 1834. By 1839, he produced both negative and positive images: photograms of lace and leaves, engravings, and photomicrographs of sliced wood and architectural views of his home. Schaaf (1992) argued that there is no evidence that Talbot told or shared his photographic work with anyone outside of his family before he became aware of Daguerre's work. Once Talbot heard news of Daguerre's photographic work, he quickly attempted to make a case for scientific priority in photography. Schaaf carefully reconstructed the British responses to the news of Daguerre's process. Once Talbot first heard of Daguerre's initial announcement in January 1839, he shared his work with fellows of the Royal Society like Sir William Jackson Hooker. Sir John Hershel would have been the logical proponent for Talbot's claims, but he had just returned from South Africa, and he was too ill to travel to London to see Talbot's work. In London for the winter and away from his laboratory, Talbot himself could not easily produce new images as evidence, according to Schaaf. But he did have with him his notebook containing a large number of images he had made in 1835. On Talbot's behalf, Michael Faraday addressed an audience of 300 at the Royal Institution on January 25, 1839, and announced the parallel discoveries of Daguerre and Talbot. Afterward, Faraday invited those present to view new library exhibits that included some of Talbot's images (Schaaf, 1992).

Talbot also moved quickly to write a nontechnical account of his methods with suggestions for photographic applications. He submitted the account, titled "Some Account of the Art of Photogenic Drawing," to the Royal Society on January 28, and a summary of it was published in the Royal Society *Proceedings*, January 31, 1839 (Schaaf, 1992, p. 51).

News that Arago announced in Paris Daguerre's success in photography on January 7, 1839, reached England almost at once. Schaaf (1992, p. 45) argued that the news of Daguerre's work likely reached Talbot by January 12 through publication in the *Literary Gazette*. Details in this first published report were few, and so initial comparisons with Talbot's own photography simply were not possible. Nevertheless, Talbot was eager to protect his intellectual work: "I was threatened with the loss of all my labour, in case M. Daguerre's process proved to be identical with mine, and in case he published it at Paris before I had time to do so in London" (Schaaf, 1992, p. 45). In fact, the details of Daguerre's process would not be disclosed until the publication of his how-to manual in August and September of 1839. As Schaaf convincingly argued, Talbot was trying to defend himself from the unknown.

Taylor (1999) argued that Talbot, during his March 21, 1839, presentation of photogenic drawing to a meeting of the Royal Society, anticipated claims for the process of *cliché verre* not only by Daguerre but also by engravers. The Graphic Society, stirred by the news from France, had assembled to discuss the implications of photography. Charles Wheatstone, a member of the Royal Society, recognized the name of Niépce in the news of the daguerreotype and sought out Franz Bauer to borrow the 1827 heliographs Niépce had left in England, to display and discuss them:

> The three key meetings of February, March, and April had allowed some the most influential figures of London's fine arts community to examine examples of processes that otherwise would have remained the province of scientists. As a result, the artists gained an understanding of the physical difference between the processes of Niépce, Daguerre and Talbot ... . The vigorous defense of the cliché verre process may have established a poor relationship between Talbot and the fine art community that never really diminished. It grew into something of a campaign against him during the early 1850s, when he was pressured into relinquishing his calotype patent. (Taylor, 1999, p. 64)

Keeler (2002) described Talbot's predicament as a reflection of "the disjunction between the older economy of landed wealth, and the economy of industrial capitalism that prevailed in England and France by the 1840s" (p. 26). Talbot obtained a patent for the process in February 1841, and, as Keeler (2002) noted, he searched for capital to realize a business vision:

> Talbot imagined training a team of artists in the calotype technique. These artists would photograph picturesque sites and monuments in the provinces. The negatives would be sent to an administrative centre, where positive copies would be made for sale and possible use in beautiful publications. (p. 26)

Eventually, Talbot sold his 10-year patent rights in France to Eugène Maret, the Marquis de Bassano. Bassano expanded the vision of the venture to include a host of photographers deployed across France who could produce albums of images that all would want to buy. Talbot, with his assistant and former valet, Nicolaas Henneman, traveled to France in 1843 to set up a photographic training workshop. When Talbot returned to England, Henneman remained for another 3 weeks. But Bassano soon redirected his energy and capital toward an even larger enterprise— developing Algerian iron ore deposits. By 1845, Bassano had ceded the French patent back to Talbot and abandoned the project altogether.

In the meantime, Talbot had come to the idea of publishing a photo-illustrated book of his own. In 1844, he established a printing workshop at Reading and made Henneman his full-fledged collaborator and manager. The first work produced was the 1844 installment of Talbot's famous book, *The Pencil of Nature*. Talbot set up Henneman in the photographic establishment at Reading, acted as the major early client in ordering prints for *The Pencil of Nature*, and later sold Henneman the entire enterprise. Although not a commercial success, the firm demonstrated the real possibility of publishing photo-illustrated texts as early as 1844. Talbot successfully responded to the same impulse for textual illustration and print duplication that first inspired Niépce. Unlike ink-based photolithographs, silver-based calotype prints tended to fade quickly due to inadequate fixing. Talbot grew frustrated with the work because of the impermanence of some of the prints (Schaaf, 1992, p. 43). He initially fixed his images with table salt, and later he fixed them with Herschel's "hypo" sodium thiosulfate. As Schaaf (1992) pointed out, Talbot, rather than giving up on photography, turned to printer's ink and to developing, as Niépce had done, a photomechanical process he termed *photoglyphic engraving* (p. 44).

As Talbot's early work with paper-based photography continued, the daguerreotype process was adopted in England, but only through the licensed agents. Elsewhere the process was not patented, and so people were free, in the United States for example, to go into business using the process without a license. The two most important early figures—Jean Francois Antoine Claudet and Richard Beard—also were competitors. Claudet learned photography from Daguerre himself, and he purchased an individual license from Daguerre. In 1841, Claudet established a studio in London. In June 1840, Beard filed a patent featuring the rival mirror camera developed by the American team of Alexander S. Wolcott and John Johnson. Beard, who used bromine to accelerate exposure times, opened England's first photographic portrait studio on March 23, 1841. Seeking a monopoly for the daguerreotype in England, on July 16, 1841, Beard worked through Daguerre's British agent Miles Berry and signed an agreement with Daguerre and Isidore Niépce to purchase the patent rights to the daguerreotype. Under the terms of British Patent No. 8194, anyone who wanted to legally engage in commercial daguerreotype in England had to apply to Richard Beard to either purchase the right of patent in a geographical area or to purchase a license to work in a specific locale. Beard and Claudet became stiff competitors, and Beard attempted to get a court injunction against Claudet, but the court found in Claudet's favor because of his individual license from Daguerre. But it is clear that Beard's defense of his patent undoubtedly deterred others from pursing daguerreotypy in England without first purchasing a license from Beard.

THE CALOTYPE AND DAGUERREOTYPE IN THE UNITED STATES—PORTRAITURE AND RADIATION THEORY

Once news of photography reached the United States, early innovators were quick to develop their own cameras, plates, and developing processes, free of patent restrictions. Two of the earliest adopters were John William Draper, an 1832 immigrant trained in chemistry at the University of London, and Samuel F. B. Morse, the painter and telegraph inventor. Like Niépce and Daguerre, this pair had somewhat differing theories of imaging. Draper's interest was primarily scientific, using photographic experiments to develop an improved theory of light and radiation. Nevertheless, his best known claim was the rather indistinct one for priority in successful human photo portraiture. A physician and a teacher, Draper was also a pioneer in lunar photography and in spectral photography; he wrote a formal theory of radiation and photomicrography. In 1840, Draper and Morse became partners in a commercial portrait studio. Morse was more aesthetic in his early theory of imaging. A visual artist by training, Morse focused on portraiture and helped to shape the aesthetic sense of such famous photographers as Matthew Brady and Albert Sands Southworth, among others. The partnership, however, was short-lived because Draper's primary interests in human photo portraiture do not appear to have been commercial.

It may very well be that Draper made the first photographic images in the United States. While a student at the new University of London from 1829 to 1832, he worked with Dr. Edward Turner and studied the chemical effects of light. After his father's death, Draper came to the United States, and in 1837, he built upon the work of Wedgwood and Davy and captured images on photosensitive materials while working toward a method of photometry. However, he was not able to fix the images. In the spring of 1839, well before the August publication of Daguerre's how-to manual, Draper read and repeated Talbot's published experiments with the paper-based photogenic drawing process. As Warner (2000) noted, Draper was able to shorten Talbot's exposure times by using a lens of large aperture and short focus (p. 16). When Draper was interviewed in 1858 about his earliest work in human portraiture, he claimed to have used Talbot's technique to create a silhouette of a person against a window, and he claimed to have photographed faces just days after the process was published in America. None of these images survived to support Draper's claim that he was one of the first to create a human portrait through photography. Unfortunately, the historical focus on daguerreotype has allowed Draper's earliest American photographs from the spring of 1839 to remain largely unresearched.

Draper returned to photographic work before the end of September 1839, just as the accounts and copies of Daguerre's process reached New York. Draper reportedly secured supplies immediately and made daguerreotypes of "brick buildings, a church, and other objects seen from my laboratory windows" the next day (Warner, 2000, p. 17). The famous portrait of his sister, Dorothy Catherine, that he sent to Sir John Herschel in 1840 was a sample of his daguerreotype portrait work (see Figure 3). As McManus (1995) showed, this portrait was clearly not the earliest, as is sometimes claimed, but more likely the product of Draper's third or fourth generation of lens improvements that he made between 1839 and 1840.

By January 1840, Draper was working with Morse, who had also been experimenting with the daguerreotype. Morse, who met with Daguerre in Paris and saw samples of the images in Daguerre's studio before it burned in 1839, wrote a

FIGURE 3   Daguerreotype portrait by John William Draper of his sister, Dorothy Catherine, 1840. As McManus (1995) showed, Draper's work in photochemistry had already led him to develop an improved camera and lens pair before the details of Daguerre's process were published in the United States. Consequently, he was among the first to make satisfactory human portraits. From the collections of the Smithsonian Institution, National Museum of American History, Information Technology and Society Division, Washington, DC. Reproduced with permission.

widely republished letter to his brother describing this new class of images. Morse resorted to superlatives as he tried to describe the daguerreotype images he first saw, writing that "the exquisite minuteness of the delineation cannot be conceived. No painting or engraving ever approached it" (Newhall, 1961, p. 15). Morse then set up a laboratory on the top of a New York University building, where he worked to increase the efficiency of the telegraph and continued to improve photography.

Draper and Morse worked collaboratively. They took in students and experimented with lighting effects, lenses, and photographic plates. For his part, Draper used photography to capture images of spectra and to theorize about spectral lines that resided outside the range of human vision—the ultraviolet and infrared (see Figure 4). This eventually led to important work in spectral photography, a field in which Draper's son Henry would excel. In addition, this work would greatly contribute to the development of astrophysics in the United States.

Elsewhere in New York, Alexander Wolcott and John Johnson designed a daguerreotype camera with a concave mirror, which gathered more light than a con-

FIGURE 4   "Solar Spectrum." A daguerreotype from 1841 by John William Draper sent to Sir John Hershel in England. Draper had been studying radiation and used photosensitive materials, including the daguerreotype plate, to record both the visible and invisible spectra. From the collections of the National Museum of Photography, Film and Television Science and Society Picture Library, Bradford, West Yorkshire, England (Image No. 10299862). Reproduced with permission.

ventional lens. In collaboration with Henry Fitz Jr., Wolcott and Johnson applied for the first U.S. patent on a photographic device, and U. S. Patent No. 1,582 was granted to Wolcott on May 8, 1840. Johnson would later successfully bring the camera design to England.

Other American photographic pioneers also were working outside of New York. In Philadelphia, Robert Cornelius, Joseph Saxton, Dr. Paul Beck Goddard, and Martin Hans Boye made and improved upon the daguerreotype process before any training from Daguerre's agents was available. Goddard appears to have been among the first to use bromine as an accelerator in the sensitizing process, which represented a major advance in shortening exposure times for portraiture. Among those who came to Philadelphia to work with them was the Englishman John Jabez Edwin Mayall. When Mayall returned to England, he worked with Claudet before purchasing a license from Beard and setting up his own London studio.

In November 1839, François Gouraud, Daguerre's agent, arrived in New York to sell cameras, plates, and training, only to find that such work had begun without him. When he arrived in Boston and Philadelphia, he met not only receptive audiences but also practicing photographers who anticipated his introduction of the daguerreotype to America. From 1839 forward, transcontinental practitioners carried training and photographic apparatus from nation to nation. What Gouraud brought to the United States, Wolcott, Johnson, and Mayall took to England. They were not only innovators but also personal agents of technology transfer who exported to England methods and devices developed in the United States.

INTERNATIONAL BORDERS AND SOME CONCLUSIONS

Photo historians have begun to look beyond some of the nationalistic rhetoric of early photography to explore how collaborations, legal relationships, and training crossed borders to widely disperse the technology in its first years. This article points to some of the key relationships, including those between Niépce and Bauer; Daguerre and Giroux; Talbot and Bayard; Claudet and Daguerre; Cornelius, Goddard, Boye, and Mayall; and Morse and Draper. What is clear is that the idea of photography, once successfully demonstrated and disseminated by Daguerre, found an appreciative audience of innovators ready to adapt the technology to a huge set of latent cultural demands for improved imaging.

As we reflect on contemporary photography, we can see how the multiple imaging theories from which early photography arose continue to serve us. In a sea of Internet graphics—graphic interchange format, portable network graphics, and joint photographic expert group format—we can recognize Niépce's initial desire for photolithography: the dissemination of graphics in texts in multiples. In cinema and high-definition television, we recognize Daguerre's drive to create the virtual reality of the diorama: the big screen with ever more verisimilitude. In every portrait studio in every city, we recognize Morse's preference for studied portraiture, Rembrandt's lighting effects, and other conventions of naturalistic European portraiture. In x-rays, magnetic resonance images, satellites, infrared images, and radio telescopes, we recognize the traces of what Draper termed the *tithonic rays* of invisible radiation. In the work and scope of international imaging corporations, Kodak or Fuji for example, we recognize the drive to supply the ever-improving apparatus, support media, and photographer training that we saw in Daguerre and Giroux, in Wolcott and Johnson, and in Talbot and Mayall. Perhaps most importantly, we recognize how long and complex a prelude the invention of photography had before Daguerre's public announcements about his process, and how quickly the process spread and was improved upon. As innovators negotiated contracts and defended patents, commercial boundaries were set, contested, and avoided with important consequences for early photo history. With close attention to the plural written histories of photography, we can resist the impulse to reduce and oversimplify the story of the origins and growth of this now fundamental modern technology and its contemporary counterparts.

REFERENCES

Batchen, G. (1997). *Burning with desire*. Cambridge, MA: MIT Press.

Bonnet, M., & Marignier, J.-L. (Eds.). (2003). Niépce, correspondance et papiers [Niépce, correspondence and papers] (2 vol.). Saint-Loup-de-Varennes, France: Maison Nicéphore Niépce.

Buerger, J. E. (1989). *French daguerreotypes*. Chicago: University of Chicago Press.

Daguerre, L. J. M. (1971). *An historical and descriptive account of the various processes of the daguerreotype and the diorama* (B. Newhall, Trans.). New York: Winter House. (Original work published 1839)

Foresta, M. A. (1996). *American photographs: The first century*. Washington, DC: National Museum of Art and the Smithsonian Institution.

Harmant, P. G. (1977). Daguerre's manual: A bibliographical enigma. *History of Photography, 1,* 79–83.

Keeler, N. (1990). Souvenirs of the invention of photography on paper: Bayard, Talbot, and the triumph of negative–positive photography. In W. Naef (Ed.), *Photography: Discovery and invention* (pp. 47–62). Malibu, CA: J. Paul Getty Museum.

Keeler, N. (2002). Inventors and entrepreneurs. *History of Photography, 26,* 26–33.

Maison Nicéphore Niépce. (n.d.). Retrieved December 9, 2005, from http://www.niepce.com/homeus.html

Marien, M. W. (1997). *Photography and its critics: A cultural history, 1839–1900*. Cambridge, England: Cambridge University Press.

Marignier, J. L. (1990, July 12). Historical light on photography. *Nature, 346,* 115.

Marignier, J.-L. (1999). Nicéphore Niépce 1765–1833: l'invention de la photographie [Nicéphore Niépce 1765–1833: The invention of photography]. Paris: Belin.

McManus, H. (1995). The most famous daguerreian portrait: Exploring the history of the Dorothy Catherine Draper daguerreotype. *The Daguerreian,* 148–171.

Naef, W. (Ed.). (1990). *Photography: Discovery and invention*. Malibu, CA: J. Paul Getty Museum.

Newhall, B. (1961). *The Daguerreotype in America*. New York: New York Graphic Society.

Rosenblum, N. (1997). *A world history of photography* (3rd ed.). New York: Abbeville Press.

Schaaf, L. J. (1992). *Out of the shadows: Herschel, Talbot, and the invention of photography*. New Haven, CT: Yale University Press.

Sontag, S. (1977). *On photography*. New York: Farrar, Straus & Giroux.

Stapp, W. F. (Ed.). (1983). *Robert Cornelius: Portraits from the dawn of photography*. Washington, DC: National Portrait Gallery.

Taylor, R. (1999). The graphic society and photography, 1839: Priority and precedence. *History of Photography, 23,* 59–67.

Trachtenberg, A. (Ed.). (1980). *Classic essays on photography*. New Haven, CT: Leete's Island Books.

Wakeman, G. (1973). *Victorian book illustration: The technical revolution*. Detroit, MI: Gale Research.

Warner, D. J. (2000). The Draper family material—National Museum of American History. *History of Photography, 24,* 16–23.

Gregory A. Wickliff is an associate professor of English at the University of North Carolina at Charlotte. He has focused his recent scholarship on the rhetoric of 19th-century photography.

TECHNICAL COMMUNICATION QUARTERLY, *15*(3), 315–328

Technology Transfer and Patents: Implications for the Production of Scientific Knowledge

Katherine T. Durack
Miami University

This article explores articulations between scientific publication and the patent system: (a) Previously patented work may function as inputs to lab activity, (b) patents may result from lab activity, (c) patents may delay scientific publication, and (d) issued patents may enhance a researcher's credibility. As patentable subject matter expands and as universities engage actively in technology transfer, researchers in cutting-edge subjects can no longer depend on pursuing inquiries in ignorance of the patent system.

In 1995, Berkenkotter and Huckin described a model of the "life cycle of lab knowledge in [the] scientific publication system" (p. 62). Published only 15 years after the passage of the Bayh-Dole Act (1980), which permitted universities to patent inventions developed under federally sponsored research programs, this model was introduced too soon to reflect the effects of patenting on the production of scientific knowledge, particularly in university environments. Central to the act of commercializing innovation, patents are the very instruments that define the ownership of technical and scientific ideas and make possible technology transfer, the "entire range of activities involved in developing new technologies and their applications to the marketplace" (Doheny-Farina, 1992, p. 3). It is the patent system that "provides a mechanism for the inventor to turn a concept into an economic value" (Bazerman, 1994, p. 81) and "to demonstrate university contributions to economic growth" (McSherry, 2001, p. 145).

As long as commercial innovation and scientific research could be considered largely separate enterprises, we might remain content with a disciplinary understanding of the production of scientific knowledge absent the potential influences of patenting. Today, however, universities and academic researchers increasingly participate in claiming and disseminating new knowledge through both journal articles and patents. Those whose research has the potential for technology transfer and commercialization must subject their work to, and negotiate the demands of, two separate systems for certifying knowledge:

- The peer review system, characterized by Berkenkotter and Huckin (1995) as "the primary means through which authority and authenticity are conferred upon scientific and scholarly papers" (p. 62).
- The patent system, the system through which the legal ownership of innovations is asserted, contested, granted, and bounded.

My goal is to describe the relationships between the patent system and the peer review system and to consider how increases in patent activity by universities can affect the production of scientific knowledge. I begin by comparing Berkenkotter and Huckin's model of the scientific publication system to the patent system. Next, I discuss four moments of articulation between the two systems that occur when academic researchers participate in technology transfer and patenting. I conclude by describing the potential effects of patenting on scientists and on the production of scientific knowledge.

SCIENTIFIC PUBLICATION AND THE PATENT SYSTEM

In the "discursive network" that Berkenkotter and Huckin (1995) described, scientific knowledge created through research activities results in two main types of textual production: grant proposals and articles for publication (p. 62). Both types of documents are subject to peer review, the "social mechanism through which a discipline's 'experts' maintain quality control over new knowledge entering the field" (pp. 62–63). The relationship between lab activities, articles for publication, and grants is recursive:

- Lab activity and knowledge result in grant proposals and articles for publication.
- Articles, once published, result in citations of the work.
- Citations enhance the credibility of the researcher and his or her knowledge claims.
- Enhanced credibility influences the persuasiveness of grant proposals.
- Accepted grant proposals fund lab activities.

The peer review process is inherently agonistic, with the editor acting as intermediary in contested ground between scientist-author and peer reviewers.

On the surface, the process by which new knowledge is authorized within the patent system bears a certain resemblance to the scientific publication system. Lab activity and knowledge of previous patents and publications (called *prior art*) also result in textual production, although within the patent system, the text produced will be either a provisional patent (explained later) or a patent application (most are utility patents; plant patents and design patents are two other options). The patent application, like an article for publication or a grant proposal, then undergoes an expert review, in this case, by an examiner in the patent office. Discussion and revision ensue if the application is rejected, as it most often is. Only when any objections have been resolved is the new knowledge authorized, the patent granted, and the technical description of the invention (in the patent document) published. Like published scientific articles, granted patents can enhance the inventor's credibility and can play an important role in securing funding to commercialize the invention or to continue other inventive activities.

There are, however, several key differences between the scientific publishing system and the patent system. Cost is one critical difference; patents can be enormously expensive to acquire and to defend. Miele (2000) estimated that "bringing a simple patent application from its first evaluation will likely cost well over $15,000," of which roughly 85% goes to pay attorney fees (p. 61). Subsequent patent litigation can cost millions of dollars (Source Translation Optimization, n.d.). Nevertheless, patents can also reap tremendous financial rewards, as Jaffe and Lerner (2004) reported:

> For numerous large companies—including, notoriously, Digital Equipment Corporation (DEC), IBM, Texas Instruments, and Wang Laboratories— ... patent enforcement activities have become a line of business in their own right. These firms have established patent licensing units, which have frequently been successful in extracting license agreements and/or past royalties from smaller rivals. For instance, Texas Instruments has in recent years netted close to $1 billion annually from patent licenses and settlements resulting from its general counsel's aggressive enforcement policy. In some years, revenue from these sources has exceeded net income from product sales. (pp. 14–15)

Potential financial gains from patenting scientific innovations make patents appealing as a source of funding for universities. *The Chronicle of Higher Education* reported in 2003 that "universities collected more than $959–million from the commercialization of drugs, software, and other academic inventions in the 2002 fiscal year" (Blumenstyk, 2003, p. A28).

The cost of acquiring patents results in the second key difference between patenting and scientific publishing: the intervention of the university in evaluating the technology and determining what is worth pursuing. Whereas a scientist's writings are largely free from university intervention, the economics of patenting result in the addition of specialized staff in technology transfer offices to evaluate inventions for patenting and to manage the process overall. Table 1 summarizes the process within the university that McSherry (2001) studied; it shows the involvement

TABLE 1
Process of Preparing a Patent Application in a University Environment

Actor	Action
University Researcher	1. Recognizes he or she has a patentable idea.
	2. Prepares a disclosure document identifying the inventors, describes the invention, and identifies potential licensees for the idea.
	3. Voluntarily submits the disclosure document to a technology transfer associate.
Technology Transfer Associate (TTA)	4. Determines who, other than the university and the researcher, might have rights in the idea. Possibilities include contributing researchers, research sponsors (public and private), and providers of tools and materials used for the research.
	5. Negotiates rights as necessary, obtaining signed agreements.
TTA + Inventor	6. Evaluates patent potential and potential interest in the idea within industry. Ideally, a licensee is identified before pursuing the patent application so costs for the patent can be deferred to the licensee.
University	7. Hires a patent attorney once the decision has been made to go forward with a patent application.
Attorney	8. Reviews disclosure and supporting materials provided by the university.
	9. Discusses the invention with the inventor.
	10. Searches for "prior art," published material covering the same or similar inventions.
	11. Prepares the patent application.
	12. Submits the application to the inventor to review for accuracy.
Inventor	13. Reviews application for accuracy, going through several iterations if necessary.
Attorney	14. Files the patent application with U.S. and (if warranted) international patent offices.

Note. The information in this table is based on information from McSherry (2001), pp. 154–157.

of the technology transfer associate (TTA) and others. Note that although the scientist-inventor prepares a disclosure document, which the TTA reviews to evaluate commercial potential, it is an attorney who actually prepares the patent application, which the scientist-inventor then reviews for accuracy. This is the third key difference: A patent document typically reflects some form of collaboration between the scientist-inventor and a legal expert.

McSherry (2001) reported that "there are no 'patent police,' " so university researchers' participation in patenting activities is largely voluntary (p. 154). Nevertheless, the practical realities of increases in patenting basic science (Rai & Eisenberg, 2003) require that researchers in certain disciplines broaden their awareness of patent literature and the potential effect of patenting on laboratory research and scientific publishing. Figure 1 shows articulations between the two discursive networks, the patent system and the scientific publication system described by Berkenkotter and Huckin (1995). To simplify this complex diagram, I do not depict the details of the technology transfer activities described by McSherry. Note, however, that the interactions McSherry described would typically occur at the intersection of the scientific publication system and the patent system denoted by *2* in the figure. In the following section, I describe articulations between the two systems.

ARTICULATIONS BETWEEN THE SCIENTIFIC PUBLISHING AND PATENT SYSTEMS

There are four moments in the life cycle of lab knowledge at which the university researcher may interact with the patent system. These moments (numbered 1–4 in Figure 1) are as follows:

1. Previously patented work may function as inputs to lab activity and knowledge.
2. Patentable inventions may result from lab activity and knowledge; thus intent to patent may affect the subsequent production of texts.
3. Patents may delay scientific publication.
4. Issued patents may enhance the researcher's credibility.

At each of these moments, patenting and patent activity can influence the knowledge-making activities of the research scientist.

Previously Patented Work as Lab Input

In the patent system, *prior art* refers to any previously known work on the technical or scientific subject at hand, including material published in a discipline's journals. The term *patented prior art* (as used in Figure 1) refers specifically to knowledge that has been the subject of an issued patent and thereby converted to private property for the specified term (20 years in the United States for the most common types of patents). Patented prior art is not restricted to the 7 million or so patents published in the United States; any information published in any international patent system would be considered patented prior art and could influence the work of researchers in a scientific laboratory. The same is true of international publications in science or technology whether academic or industrial in origin.

The subject matter eligible for patent protection varies worldwide. In the United States patentable subject matter of primary interest includes "any new and useful process, machine, manufacture, or composition of matter," which are covered by utility patents (Patentability of Inventions, 2001). Laboratory equipment, procedures, and materials may receive utility patents; examples include

- An apparatus used for electrophoresis (Hunter, Rummery, Herbert, & Durack, 2005).

Patent System

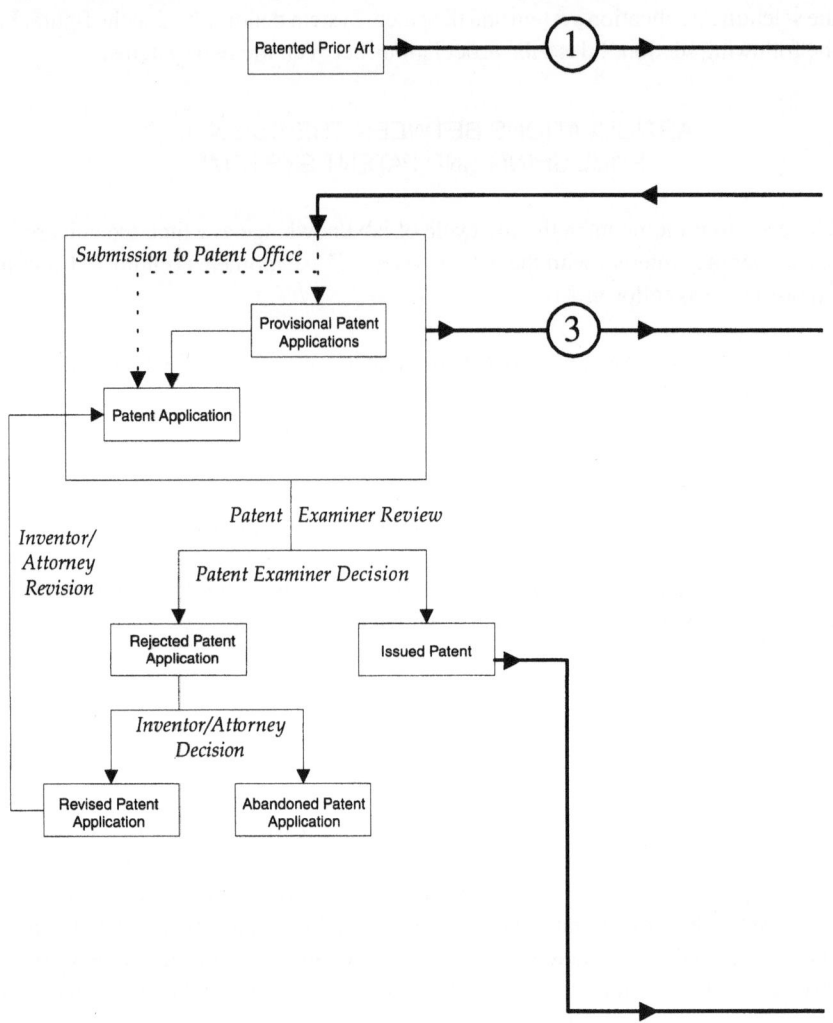

FIGURE 1 Articulations between the discursive activities of the patent system and the scientific publishing system. Portions of this figure are from *Genre Knowledge in Disciplinary*

(*continued*)

Scientific Publishing System

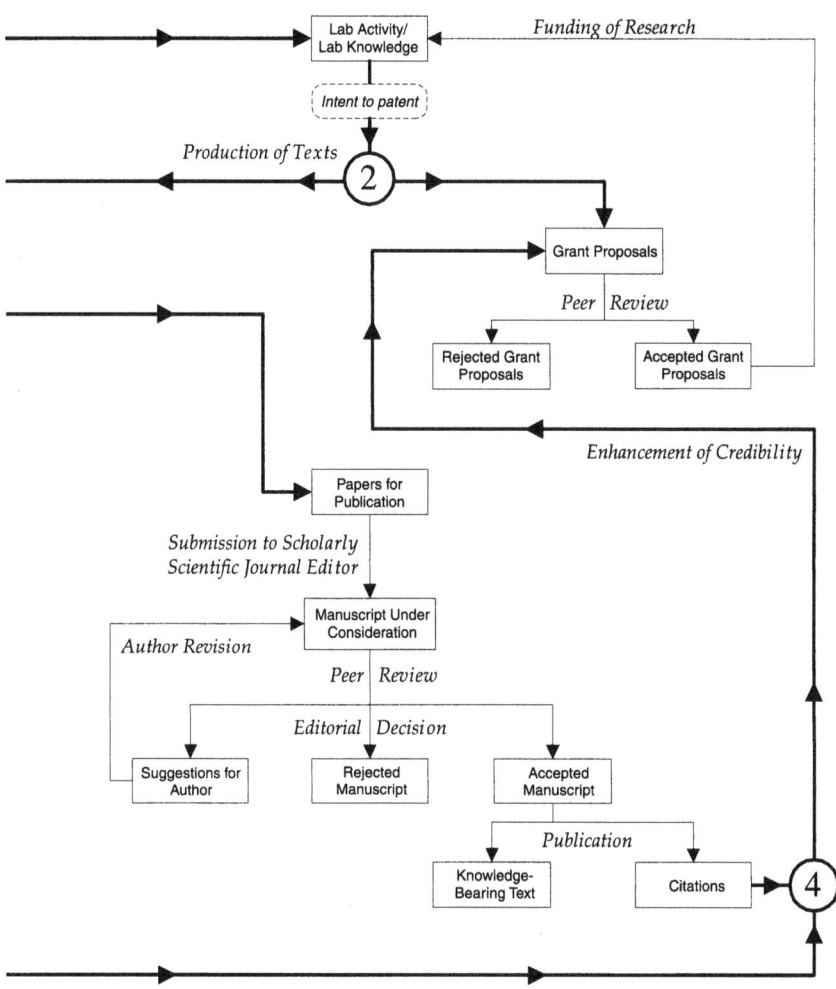

Communication: Cognition, Culture, Power (p. 62), by C. Berkenkotter and T. N. Huckin, 1995, Hillsdale, NJ: Lawrence Erlbaum Associates, Inc. © 1999 by Lawrence Erlbaum Associates, Inc. Adapted with permission.

- Cell-sorting equipment and a method for providing sex-sorted animal sperm (Durack et al., 2005).
- Genetically engineered bacteria used to manufacture some types of drugs (Rai & Eisenberg, 2003).
- A genetically engineered mouse (the *oncomouse*) patented by Harvard University, used for cancer research (Rai & Eisenberg, 2003).
- Embryonic stem cells for all primates, patented by the Wisconsin Alumni Research Foundation (Rai & Eisenberg, 2003).

Other inventions patented by universities that are of recent note include predictive text technology for entering text messages using a standard telephone keypad, assigned to the University of Texas at Austin (Mangan, 2005); antipiracy software to combat illegal online music trading, assigned to the University of Tulsa (Read, 2004); and Web browser technology owned by the University of California (Foster, 2005).

Finally, the information in patent documents may prove useful to researchers. Walker (1995) asserted that "patent documents remain the only source for some of the most useful and significant records of recent discovery, development and refinement of research in many fields, information of both economic importance and scientific significance" (p. 1). The value of patent documents to researchers in a given field undoubtedly varies; in addition (as I will discuss later), the rhetorical peculiarities of the patent system can compromise the usefulness of patents as a source of technical and scientific information, even though disseminating the information in patent documents is one key purpose of the patent system.

Patents as a Result of Lab Activity and Knowledge, and an Influence on the Timing of Scientific Publication

University researchers may achieve patentable outcomes as a result of their scientific investigations. When this is anticipated, university researchers must be cognizant that patent laws restrict what may be published before an application is filed. U. S. law allows a one-year grace period after public disclosure before an application must be filed. If the commercial viability of an invention is uncertain, an inventor may also choose to file a provisional patent application instead of a regular patent application. (A provisional patent application permits the researcher to obtain an early filing date at a much lower cost and without meeting all of the requirements of nonprovisional patent applications.) In contrast, international patent laws make no allowance for public circulation of knowledge before the application is filed, requiring *absolute novelty*. Absolute novelty "means that the invention must not have been previously known to the public, sold or offered for sale (publicly or secretly), or made available to the public anywhere in the world in any way before the filing date" (Copeland & Meagher, 2005).

The novelty requirement, both within the United States and abroad, requires researchers to take special care when communicating about their work. To preserve foreign rights, patent applications should precede—or at least coincide with—journal publication, possibly delaying the publication of scientific journal articles. Yet the effect of the novelty requirement for patent applications extends even further: Less obvious is the potential influence of patenting on other forms scientific communication. Grant proposals have been deemed sufficiently public to be considered publications in the eyes of the law. Unless special precautions are taken when writing a proposal (by identifying potentially patentable material and marking it as confidential), a researcher may unwittingly start the clock on the U.S. grace period and, if a patent application is not submitted within 1 year, may lose rights to potentially patentable subject matter mentioned in the proposal (University of Cincinnati Intellectual Property Office, n.d.). Yet even these precautions might fail, depending on how the applications are handled by the funding agency. The National Institutes of Health, for instance, recently posted confidential applications on the Web in error, an act that could compromise the ability of researchers to protect patentable outcomes (Pulley, 2005). Conference presentations and doctoral dissertations also have been considered sufficiently public to bar patenting (Copeland & Meagher, 2005).

Patents as an Enhancement to a Researcher's Credibility

Like scientific publications, patents may enhance the researcher's credibility, thereby improving the persuasiveness of grant proposals. Yet within the scientific community, patents are not universally granted the same stature as journal publications. Although McSherry (2001) indicated that some of her respondents took great pride in their patents, finding them "a validation of creativity" (p. 165), she also reported that "scientists believe patents obscure and/or ignore method and are therefore illegitimate vehicles for scientific knowledge" (p. 175). She quoted one senior researcher as remarking, "I have to worry about my papers. That's more important in my career than having patents" (p. 187).

With whom, then, does an issued patent enhance credibility? In the context of producing texts to secure funding, the philosophy is that patents establish credibility with, and provide protection to, private investors who otherwise would be unwilling to cover the costs associated with bringing an invention to market. In addition, new inventions may be "given away" by faculty researchers "in exchange for research funding and consulting fees and/or to use university facilities for commercial purposes," which is an unacceptable practice that violates typical university policies regarding patents (McSherry, 2001, p. 158). Given the express desire to see public benefits derive from federally funded research, we might presume that federal agencies would be favorably influenced by a history of patenting and technology transfer. Outside the context of external funding, patents may favorably influence promotion and tenure committees as well (McSherry, 2001).

ISSUES IN PATENTING AND SCIENTIFIC PUBLICATION

As universities become more actively engaged in patenting the work of academic researchers, the experience of scientific publication and patenting by researchers in industry provides a cautionary message about relationships between the two systems. Drahos and Braithwaite (2002) were critical in reporting DuPont's practices in the early 20th century. After a scientist at DuPont published an article that undermined one of the company's nylon patents, thus allowing a competitor to take advantage of the technology, DuPont tightened its controls over researchers' communications. According to Drahos and Braithwaite,

> The upshot was that some scientific articles did not get published. DuPont began to get a reputation among the general scientific community for feeding off the research efforts of others without returning anything to the community. Its obsession with patent protection conferred upon it the reputation of being a free-rider. (p. 45)

More recently, attorneys writing in *The Chronicle of Higher Education* advised readers that "the lesson for faculty members is clear: A seemingly innocuous action like submitting a thesis or posting research results may start the countdown for filing a patent application" (Copeland & Meagher, 2005).

It is no surprise that some fear the trend toward seeking intellectual property protection for university research will damage the culture of open exchange that science has claimed as its hallmark. Drahos and Braithwaite (2002) observed that "the university itself ... has been the greatest fount of innovation," (p. 211) yet they cited increasing university involvement in patenting as evidence that the current intellectual property environment is one of "information feudalism," in which knowledge no longer comes to light as a public good (p. 218). Intent on addressing such concerns, a special investigation carried out by the Commission of the European Communities (2002) recently studied whether patenting impeded the publication of scientific research in biotechnology. The results were in many ways encouraging, showing that scientists familiar with both systems can participate in each and perceive minimal delays in publishing scientific journal articles (p. 12). Nevertheless, academics who responded to the survey identified having a grace period (such as that offered by provisional patents in the U.S. system) as "the most important measure to minimise delays in scientific publication" (p. 13).

Publication delays are not the only concern. Drahos and Braithwaite (2002) are even more extreme in their charge that "scientists increasingly are vassals of knowledge corporations ... [whose] freedom of inquiry is blocked at various turns by patent and copyright obstacles" (p. 201). In one specific example, medical researchers studying cancer "have been forced to abandon their research programs due to licensing terms" because breast cancer genes useful for that research are privately owned (Jaffe & Lerner, 2004, p. 17). Furthermore, as they adopt corporate-type practices, such as patenting the inventions of research scientists, universities may increasingly be viewed by the courts as businesses, thus limiting the extent to which they can be exempt from infringement—an exemption universities have formerly enjoyed because of their nonprofit status (Blumenstyk, 2002). To shed light on whether "academe's growing pursuit of patents, and related practices for commercializing research, 'are aiding or hindering the progress of science,' " the American Association for the Advancement of Science announced in March 2005 its intent to survey[1] some 4,000 scientists, including 1,000 scientists overseas (Blumenstyk, 2005). Another study sponsored by the same organization investigates the extent to which academic institutions have been receiving infringement notifications (notifications that claim unlawful use of patented subject matter; Blumenstyk, 2005, p. A31).

On a more fundamental level—and perhaps, of greater interest to writing researchers—is whether patents actually meet the statutory requirement of disseminating technical and scientific knowledge to the public. It is to ensure the sharing of technical and scientific knowledge that the system *propertizes* such knowledge in the first place, providing in exchange a short-term monopoly as an incentive. Though intended to be a transparent disclosure of technical and scientific fact— and despite surface similarities to the scientific journal article—"patent applications are always arguments, bound by a set of contested rhetorical rules" (McSherry, 2001, p. 169). At the heart of this contest for meaning is a fundamental tension between legal discourse and scientific discourse. Legal discourse is "held accountable … to a hierarchically arranged series of court decisions, laws, and constitutions, and … to evidence gathered through procedures defined by the system and represented in a manner established by tradition and explicit rule" (Bazerman, 1988, p. 61). In contrast, scientific discourse is "built on accountability to empirical fact (as … characterized within the thought style of science) over all other possible accountabilities" (Bazerman, 1988, p. 62). Drahos and Braithwaite (2002) are critical of the resulting obfuscation in patent, writing

> Drafting patent applications [has] developed into a special kind of art. Since knowledge was the basis of competitive advantage, it followed for all companies that they should disclose as little of their knowledge as possible. But the patent system required the disclosure of the invention to the public. … [One] solution to the problem of public disclosure was a drafting one. Patents were drafted in ways that satisfied the patent office, but were virtually useless to public readers of the documents. The best

---

[1]This study was released in October 2005 (see Hansen, Brewster, & Asher, 2005).

patent attorneys took the art of the "empty" but valid patent specification to spectacular heights. (p. 47)

As McSherry (2001) observed, "for the attorney, the object is not to lie but rather to include *all* possible truths. For the researcher, the process may seem either so invasive or so much sophistry, and in any case an affront to some traditional academic principles" (p. 173, emphasis in original).

This tension between legal rules and scientific facts played out in the pages of *World Patent Information* (Ustinova & Chelisheva, 1996) in which chemists took issue with the way legal discourse has distorted the information in chemical patents. The authors assert that "chemists do not want 'paper' chemistry but patents that disclose real chemical compounds rather than graphic images of imaginary molecules" (p. 29). More recently, Stuart Newman's efforts to patent a human–animal chimera exemplify tensions between the progress of science and the business of patents (Dowie, 2005, p. 91). Newman submitted his patent as a way to make a statement about the morality of owning human life in cases where, for example, a chimpanzee embryo might be injected with human embryonic stem cells, resulting in a transgenic animal that could be used for drug research or to produce human-transplantable organs (Dowie, 2005, p. 96). Such transgenic animals currently exist: The oncomouse (mentioned earlier) is one example, as is the *geep*, which includes genetic material from goats and sheep (Dowie, 2005, pp. 95–96). Newman submitted his application to force the U.S. Patent and Trademark Office to make a ruling about ownership of human life, figuring that either outcome—a rejected application or a granted patent—would advance his quest to require the scientific community to address questions of personhood, such as whether "one animal cell make[s] a being suitable for ownership, forced labor, and medical experimentation, just as 'one drop' of black blood once did" (Dowie, 2005, p. 102). After 7 years of effort, Newman's patent application was denied in February 2005 (Weiss, 2005).

CONCLUSION

Critics proclaim that the patent system is "broken" and is "endangering innovation and progress": "The intense pathology of the current system arises from the combination of stronger patent protection, a decline in standards for granting patents, and the emergence of broad, apparently invalid, patents in particular industries undergoing rapid technological change" (Jaffe & Lerner, 2004, p. 19). Despite these problems, the U.S. patent system is credited with playing a key role in the evolution of the country "from a colonial backwater to become the pre-eminent economic and technological power of the world" (Jaffe & Lerner, 2004, p. 1). If, as Walker (1995) claimed, "less than ten percent of the information appearing in patent documents is ever published elsewhere" (p. 4), then patents have enormous value as a source of technical and scientific information even though their complex rhetorical structures may inhibit clear communication. In any case, increases in patenting have "blur[red] traditional distinctions between mechanisms for disseminating basic research findings and applied inventions" (Hansen et al., 2005, p. 6). As patentable subject matter expands and as universities engage more actively in technology transfer, researchers will need to become adept at managing the relationships between knowledge validation in the patent system and the scientific publishing system. Indeed, it is becoming increasingly clear that researchers in cutting-edge subjects can no longer depend on having the ability to pursue scientific inquiries in ignorance of the patent system.

ACKNOWLEDGMENTS

I am grateful to Paul Anderson and David Marado for their suggestions on Figure 1 and to Laura Howard for bringing Dowie's essay, "Gods and Monsters," to my attention. Sue and Tom Shay reviewed an early version of this article and provided essential feedback. A portion of the research described in this article was completed during an Assigned Research Appointment granted by Miami University during Spring 2004; I am grateful to the Miami University Committee on Faculty Research for its support.

REFERENCES

Bayh-Dole Act of 1980, 35 U.S.C.S. § 200 *et seq.*

Bazerman, C. (1988). *Shaping written knowledge: The genre and activity of the experimental article in science*. Madison: University of Wisconsin Press.

Bazerman, C. (1994). Systems of genres and the enactment of social intentions. In A. Freedman & P. Medway (Eds.), *Genre and the new rhetoric* (pp. 79–101). London: Taylor & Francis.

Berkenkotter, C., & Huckin, T. N. (1995). *Genre knowledge in disciplinary communication: Cognition, culture, power*. Hillsdale, NJ: Lawrence Erlbaum Associates, Inc.

Blumenstyk, G. (2002, October 25). Court ruling revives patent lawsuit against Duke U. *Chronicle of Higher Education,* A33.

Blumenstyk, G. (2003, December 19). Inventions produced almost $1-billion for universities in 2002. *Chronicle of Higher Education,* A28.

Blumenstyk, G. (2005, March 4). Science association assesses impact of quickening drive for patents. *Chronicle of Higher Education,* A31.

Commission of the European Communities. (2002). *Report from the Commission to the European Parliament and Councill—An assessment of the implications for basic genetic engineering research of failure to publish, or late publication of, papers on subjects which could be patentable as required under Article 16(b) of Directive 98/44/EC on the legal protection of biotechnological inventions— COM (2002).* Retrieved September 28, 2005, from http://europa.eu.int/comm/internal_market/en/indprop/invent/index.htm

Copeland, R. G., & Meagher, T. F. (2005, May 27). Academics must protect their patent rights. *Chronicle of Higher Education,* B6.

Doheny-Farina, S. (1992). *Rhetoric, innovation, technology: Case studies of technical communication in technology transfers.* Cambridge, MA: MIT Press.

Dowie, M. (2005). Gods and monsters. In A. Lightman (Ed.), *The best American science writing* (pp. 90–102). New York: Harper Perennial.

Drahos, P., & Braithwaite, J. (2002). *Information feudalism: Who owns the knowledge economy?* New York: New Press.

Durack, G., Wallace, J. D., Vandre, G. P., Westfall, L. A., Hatcher, J. T., & Nayak, N. V. (2005). *U.S. Patent Application No. 20050112541*. Washington, DC: U.S. Patent and Trademark Office.

Foster, A. L. (2005, September 30). U.S. patent office upholds web-browser patent issued to U. of California and disputed by Microsoft. *Chronicle of Higher Education Daily News Archive*. Retrieved December 20, 2005, from http://chronicle.com/daily/2005/09/2005093001t.htm

Hansen, S., Brewster, A., & Asher, J. (2005). *Intellectual property in the AAAS scientific community: A descriptive analysis of the results of a pilot survey on the effects of patenting in science*. Retrieved November 1, 2005, from American Association for the Advancement of Science, Directorate for Science and Policy Programs website: http://sippi.aaas.org

Hunter, W., Rummery, G. M., Herbert, B. R., & Durack, M. C. A. (2005). *U.S. Patent Application No. 20050072678*. Washington, DC: U.S. Patent and Trademark Office.

Jaffe, A. B., & Lerner, J. (2004). *Innovation and its discontents: How our broken patent system is endangering innovation and progress, and what to do about it*. Princeton, NJ: Princeton University Press.

Mangan, K. S. (2005, April 1). U. of Texas files patent lawsuit. *Chronicle of Higher Education*, A39.

McSherry, C. (2001). *Who owns academic work? Battling for control of intellectual property*. Cambridge, MA: Harvard University Press.

Miele, A. L. (2000). *Patent strategy: The manager's guide to profiting from patent portfolios*. New York: Wiley.

Patentability of Inventions, 35 Part II U.S.C. § 10-101. (Washington, DC: U.S. Government Printing Office 2001).

Pulley, J. L. (2005, October 31). NIH accidentally posts confidential grant applications on the web. *Chronicle of Higher Education Daily News Archive*. Retrieved December 20, 2005, from http://chronicle.com/daily/2005/10/2005103103n.htm

Rai, A. K., & Eisenberg, R. S. (2003). Bayh-Dole reform and the progress of biomedicine. *American Scientist 91*(1), 52–59.

Read, B. (2004). New software uses fake songs to confound would-be music traders. *Chronicle of Higher Education, 50*, A25. Retrieved December 20, 2005, from http://chronicle.com/cgi2-bin/texis/chronicle/search

Source Translation Optimization. (n.d.). *Legal resources and tools for surviving the patenting frenzy of the Internet, bioinformatics, and electronic commerce*. Retrieved December 20, 2005, from http://www.bustpatents.com

University of Cincinnati Intellectual Property Office. (n.d.). *How to prevent grant proposals from barring patent protection*. Retrieved October 1, 2005, from http://www.ipo.uc.edu/index.cfm?fuseaction=overview.how

Ustinova, E. A., & Chelisheva, O. V. (1996). Are Markush structures matters of chemistry and law or just figments of the imagination? *World Patent Information, 18*, 23–31.

Walker, R. D. (1995). *Patents as scientific and technical literature*. Metuchen, NJ: Scarecrow Press.

Weiss, R. (2005, February 13). US denies patent for part-human hybrid: Scientist aimed to prevent others' use. *Boston Globe*. Retrieved October 28, 2005, from http://www.boston.com/news/nation/washington/articles/2005/02/13/us_denies_patent_for_part_human_hybrid?pg=2

Before joining the faculty at Miami University, Katherine Durack worked as a writer, manager, and consultant in information technology. She has a long-standing interest in intellectual property and patent documents as technical and scientific texts.

TECHNICAL COMMUNICATION QUARTERLY, *15*(3), 329–353

Social Determinants of Preparing a Cyber-Infrastructure Innovation for Diffusion

Barbara Mirel
University of Michigan

Nicholas Johnson
University of Michigan

This study presents a case of asynchronous, collaborative problem solving aimed at readying a sophisticated distributed technology for large-scale diffusion. We analyzed e-mail transcripts of 30 technologists negotiating complex technical improvements necessary for wide-scale diffusion and found that the group's social interactions and discursive practices determined the improvements they were willing to realize. We detail these social dynamics and their effects on readying technologies for diffusion and argue that technology teams need to become more aware of diffusion as a social dynamic.

Rolling out a new Internet is a major undertaking, and technologists have been preparing this technology transfer for the past 8 years. Internet2 (I-2) is a high-performance, high-speed next-generation Internet designed for the needs of remote scientific collaborations, distance education, and health care. Its fat pipes (bandwidth enough for all media) and unprecedented speeds (the equivalent of 57,000 pages of text/sec) enable people to exchange huge volumes of information quickly and reliably.

But bandwidth and speed alone cannot guarantee successful diffusion of this innovative technology. As current pilot testing in the field shows, I-2 still performs unacceptably often enough to warrant holding off on mass transfer until substantial improvements are achieved. One such improvement relates to the persistent problem of systemic packet loss, transactions that are lost or dropped as they move from sender to receiver across diverse network domains and edges. They are lost not because of congestion but because of system incompatibilities.

Our study focused on I-2 specialists' attempts to generate needed improvements for systemic packet loss to ready I-2 for technology transfer. These efforts are intrinsic to technology transfer because they shape the design of a deployed technology and its acceptance (or not) by users. We examined a case of 30 I-2 technologists from diverse network domains who collaborated remotely via e-mail for 13 consecutive days to examine and resolve root causes of systemic packet loss and associated minor problems. We analyzed their e-mail transcripts for sociocommunicative factors affecting success in realizing technical improvements.

Our study revealed that in this distributed, decentralized sociotechnical system, social forces have more influence than technical ones in preventing multidomain technologists from realizing some of the harder redesign improvements required for readying their innovation for diffusion. In our case, the participants did not realize that their social moves limited the technical improvements they were willing to consider and, at times, adversely affected their ability to decide on designs needed for improved network performance. For decentralized settings that are technically and socially networked, our case identified variables that can be used to indicate critical social dynamics that determine and, at times, obstruct technological improvements necessary for diffusion. We argue that teams need to be more aware of these dynamics to move forward in designing technical innovations to ready infrastructures for wide-scale diffusion.

DESCRIPTION OF THE EXAMINED PROBLEM

In October 2002, a CEO's hour-long presentation during a nationwide videoconference sporadically lost audio and video reception at remote sites, and the problems were not solved until after the presentation ended. Troubleshooters had difficulty finding and resolving the problem because different tests gave contradictory readings about network performance. One test showed no packet loss, another measured 1% loss, and a third indicated 20% loss. They also saw nothing untoward posted on the Weathermap, the name given the network-wide display of known problems. Initially, local network operators tried solving the problem themselves, but after a fruitless hour they sent a message to the I-2 e-mail list of end-to-end (e2e) specialists asking for help. Several specialists responded, and working jointly with local network operators, finally identified the source of the error two hours later. Unreported repairs of optical gear in one part of the network had triggered problems in protocol and application responses at the edges of another network. After finding the problem, the collaborators rehashed the event and its causes until 11 p.m. They determined it was a compound problem of knowledge management and systemic packet loss. First, in terms of knowledge, information about the equipment repair was not posted on the Weathermap. Second, systemically incompatible network configurations and application designs prompted adverse responses from the application. By the next day, more people from the e2e mail list had joined in. Ultimately, 30 people participated via 102 e-mails.

SOCIOTECHNICAL CONTEXT

I-2 started as a response to the commodity Internet, commercialized in 1995 but deemed too slow and congested for the long-distance, media-rich exchanges and simultaneity required in such areas as scientific research, distance education, and medicine. As an alternative, I-2 was developed and intended for mass diffusion.

Understanding I-2 as an Organization

I-2 is a community and a technology. As a community, it is a consortium of more than 200 research universities, various governmental partners, and corporate sponsors who share the goal of creating and using a high-performance network. Stakeholders also include network operators and engineers who work in the campus, local, regional, and national networking infrastructures that connect to the I-2 backbone.

I-2 is a distributed and decentralized organization of groups that design, develop, improve, maintain, and administer this innovative infrastructure. These groups are highly autonomous yet highly interdependent. Each oversees its respective portion of the system and works with others to assure acceptable performance and solve problems. The community values consensus and coordination, with both dynamics essential for solving complex problems and instituting improvements.

Because complex problems affect numerous subnetworks and elements and are often cascading, emergent, and hard to isolate, I-2 created the e2e performance group in 2001. It is charged with improving system incompatibilities and other e2e issues. Staffed by eight experts in the integrated workings of the technology, this group has four objectives: (a) developing an e2e performance-measurement infrastructure, (b) developing e2e problem-solving tools and knowledge bases, (c) specifying necessary designs for bringing application and network processes in synch, and (d) fostering relevant communications and coordination between diverse groups.

For the past several years, the e2e group has predominantly concentrated on the first two objectives just mentioned—constructing a measurement infrastructure and developing tools and information. But, as findings from this study show, the last two objectives—design and communication—must be better emphasized. Diagnostic tools and documented information cannot substitute for redesigns that bring application and network processes in synch. For these redesigns, more than for any other improvement, realization requires optimal communications and coordination.

Understanding I-2 as a Technology

Technologically, the I-2 infrastructure comprises the network backbone, called Abilene; a wide range of network applications and applications to facilitate ongoing scientific research projects; workstation capabilities; and campus, local, regional, and national infrastructures that enable high-performance network connections. Some connector networks feed into Abilene directly, whereas others link to additional connector networks. I-2 also includes the hardware, services, transport protocols, and tools that enable the transmission and support of communications. Based on capacity alone, Abilene can transfer data at 2.5 gigabits/sec, so, as previously mentioned, congestion is not a problem. However, applications for audio, video, real-time data streams, and synchronous participation in virtual environments have different transmission requirements. As a result, these applications demand intersystem flexibilities and compatibilities that I-2 specialists are still learning about.

RELEVANT LITERATURE

Our case was an instance of distributed and decentralized computer-supported collaborative work (CSCW), but its distinct traits have rarely been covered as a whole in either the CSCW literature or other areas of study. Specifically, our case involved cross-domain temporary teams in a decentralized organization collaboratively solving an unanticipated complex problem, examining its root causes, and determining remedies that could prepare the technology for diffusion and user acceptance.

The research literature has addressed in isolation some of the sociotechnical dynamics and effects relevant to our case, but we argue that these traits must be studied as a set because as a set they characterize growing numbers of decentralized, networked organizations and problem-solving collaborations. Given this qualification, the literature relevant to our case falls into three areas of inquiry that mirror the three analytical approaches we took to study the I-2 e-mail transcripts, namely (a) the logic and arguments people use to explore and debate complex problems; (b) the social dynamics associated with solving complex problems cooperatively and remotely when teams are functionally diverse, decentralized, or temporary; and (c) motivations that prompt people to pursue less than optimal choices in complex problem solving. Before presenting our questions and analysis in these areas, we turn to insights from studies examining similar issues in similar ways.

Logical Arguments in Complex Problem Solving

Researchers who focus on problem solvers' argumentative structures in complex problem solving often have used a special form of conceptual mapping called cognitive mapping. Findings from such studies have shown that, contrary to common assumptions, domain experts such as our I-2 technologists do not oversimplify situations through such strategies as "satisficing" (pursuing the first good plan) or attending selectively (artificially) to only one goal when many goals at once are operative (Axelrod, 1976; Wellman, 1994). Instead they strive to turn complexity into stable chains of causal relationships. Unfortunately, they find it hard to achieve this stability when the issues at hand are ambiguous, involving, for example, uncertain means–ends relationships or competing goals. Sociologically, researchers have shown that problem solvers deal with such ambiguities by discussing contested issues at an abstract or thematic level, thereby avoiding a detailed examination of premises and assumptions that may divide the group and thus disrupt harmony (Weick, 1995). This strategy is especially common in situations requiring coordination among decentralized, diverse groups, as in our case study. But complex problem solving can succeed only if people mix abstract with detailed reasoning about goals, means–ends relations, and responsibilities (Ham & Yoon, 2001; Schaafstal, Schraagen, & van Berlo, 2000). Therefore, in the intricate web of fleshing out issues and arguments in complex problem solving, people's methods for assuring social harmony and integration may be at odds with the methods best for sorting out, solving, and taking action on root causes of complicated problems. This was the issue in our case study.

Social Dynamics Specific to the Traits of Our Case Study Group

A good deal of research has shed light on social dynamics in arguments and problem solving that aim at constructing common ground in virtual team collaborations while still resolving thorny and politically charged technical problems. Our case of I-2 problem solving took place virtually and asynchronously via e-mail. Studies such as Birnholz, Finholt, Horn, and Bae (2003) have established that people can succeed in creating common ground through cool media like e-mail, but unlike in our I-2 case, Birnholz et al.'s findings relate to exchanges with informational content and little if any ambiguity (Kramer, Fussell, & Setlock, 2004; Postmes, Spears, & Lea, 2000; Tidwell & Walther, 2002; Walther, 2004). Unfortunately, few if any studies on e-mail collaborations have examined exchanges amid time pressures about complex problems that involve ambiguous issues, technical and social relationships, and boundaries, as our study did.

Generally, research on remote collaborations has determined that the distance implicit in this collaborative problem solving creates difficulties that require participants to attend closely to social dynamics so that they exchange and explain the right types and amounts of knowledge. As a means to this end, several studies have emphasized the need for trust (Herbsleb, Mockus, Finholt, & Grinter, 2001; Jarvenpaa & Leidner, 1999; Panteli & Sockalingam, 2005). Virtual temporary teams, however, do not have sustained time together to construct this trust. There-fore, they typically develop what Meyerson, Weick, and Kramer (1996) called "swift trust" by dealing with one another in terms of their roles instead of individu-ality. Role clarity helps temporary teams like ours develop common grounding and successful communications (Hollinghead, 1998). Two other factors that contribute to swift trust are (a) explicitly exploring diverse expectations about how team-mates should accomplish tasks and allocate responsibilities, and (b) explicitly ex-pressing affective and interactive stances and motivations, which sustain high lev-els of activity (Coppola, Hiltz, & Rotter, 2004).

Research has shown that it is not enough for remote collaborators to enact ef-fective group processes. They also must be aware of them. For hard problems, this awareness is as important as teammates' intellectual understanding of the problem (O'Rorke & Henrion, 1996). But this awareness is often elusive when cross-functional participants like those in our study work jointly on highly equivo-cal inquiry tasks with mixed priorities (Jarvenpaa & Leidner, 1999). In these situa-tions, they often avoid potential discord, which mutes overt comments about group processes. Motives here become mixed. For complex problem solving, as Panteli and Sockalingam (2005) argued, it is vital for teammates to deal overtly with dis-cord rather than avoid it because conflict, when handled well, is an invaluable cre-ative spark for innovative thinking and novel options. We addressed this issue of conflict and problem solving in our study by examining in the e-mails indicators of agreement, disagreement, trust, role-based stances, and explicit expressions about collaboration. We also looked for indicators of social dynamics that are distinct to decentralized organizations, such as delegating tasks through a "first come–first served strategy" (Malone & Crowston, 1994). In this strategy, teammates do not explicitly assign tasks and responsibilities to each other but instead tacitly expect whoever is motivated to take up a suggested solution, and the first feasible out-come is implemented.

Motivations for Social and Discursive Choices

Knowing that certain social dynamics occur in collaborative problem solving is in-complete without understanding the motives that may account for choices partici-pants make and their consequences. For example, Weick (1995) contended that cross-functional teammates are apt to make choices counterproductive to coopera-tive problem solving if they are thrown unexpectedly into intense sensemaking—the situation our participants were in. In these situations, people's first inclination may be self-protection, and they will assume a stance that avoids dependency on others. Similarly, when multifunctional collaborators engage in complex intellec-tual exchanges, role clarity often diminishes because individuals flexibly assume several identities in regard to various claims. Their interactions "often look more like complex exercises in social construction of identity than like the mechanical spread of information" (p. 102). Moreover, teammates' motivations to establish group perceptions of similarity may explain exchanges that oversimplify situations and solutions or avoid difficult or ambiguous issues (Strang & Meyer, 1994). In sum, motives for choices people make in collaborative problem solving are mixed, making it hard to achieve at once social integration, individual autonomy, and solutions to complex problems and conflicts.

As Burke (1969) emphasized, from this perspective exploring unclear relations and ambiguous subjects in complex problems is not simply an exercise in logic. It is a communication act, with significant consequences issuing from people's choices about how to say what they think, to whom to say it, and when. Burke's dramatistic analysis of written texts is a means for uncovering motives for these choices. In our analysis of e-mails, we found, as Burke and others (Malone & Crowston, 1994) had, that by examining how individuals discursively negotiate ambiguities, analysts can uncover the best opportunities for change and improvement

METHOD

Drawing from current research, we analyzed the problem-solving messages in the I-2 transcript using three complementary methods: (a) cognitive mapping, (b) content analysis, and (c) dramatistic analysis. When we began the study, we intended to use cognitive mapping to clarify and structure the e-mail exchanges in terms of goal-driven causal reasoning (inherent in any root-cause analysis) and to use content analysis to gain a more thorough account of the social realities affecting participants' emphases on certain themes, dispositions, and networks of interactions. We determined to do a rhetorical analysis only after we began analyzing results from these two methods because we could not tell from those results why participants interacted and conversed as they did. Dramatistic analysis would provide these insights. The dramatistic analysis, which proved to be invaluable, helped us validate and extend our findings from cognitive mapping and content analysis.

Cognitive Mapping

Cognitive mapping is a means for modeling the structure of writers' arguments into complex chains of causal–inferential reasoning that aim to achieve desired goods or utilities (Axelrod, 1976; Wellman, 1994). Typically, cognitive mapping is applied to a single author's text. However, we adapted it to an analysis of all postmortem e-mail messages as a single transcript to better understand the group as a semantic community. In this way, we examined the scope and reasoning used by these I-2 collaborators to relate premises, implications, and outcomes to utilities. We omitted troubleshooting exchanges from the first day because they were not arguments. We examined (a) the scope of the arguments problem solvers mutually agreed on, (b) the ways problem solvers subdivided problems and goals to make them manageable, (c) arguments that converged and ran counter to each other, and (d) the extent to which the participants completely and complexly dealt with the issues.

This methodology differs from representing causes, effects, and beliefs in an Ishikawa (fishbone) diagram, a common method of root-cause analysis for quality assurance (Stålhane, Dingsøyr, Hanssen, & Moe, 2003). Like fishbone diagrams, however, cognitive maps can reveal how comprehensively participants explore the full range of potential root causes, including, for example, an analysis of methods, resources, people, environment, tools, and training.

In constructing the cognitive map, we followed procedures developed for cognitive mapping in Wrightson's (1976) "Documentary Coding Method" and in Wellman (1994)—comprehensive and detailed procedures for identifying and coding assumptions, cause and effect concepts, independent and dependent relationships, different types of causal relationships, and linkages that positively or adversely affect utilities. Cognitive maps do not reflect the temporal order of writers' assertions and typically do not include reliability measures. To analyze the resulting cognitive map, we identified when and why participants used contingent reasoning, inferences, and trade-offs, and what causal and conditional logic they used to link problem traits, improvements, and utilities. We also compared findings to an analysis of numerous I-2 historical documents to see whether arguments and improvements in this case were new news. Our cognitive maps revealed the structure and omissions in participants' arguments but not the relative importance they placed on specific themes, dispositions, and interactions. To gain these insights, we conducted a content analysis.

Content Analysis

Our content analysis substantively examined what the I-2 collaborators agreed upon and debated and what social and communicative processes they used to construct shared knowledge, role identities, and group ethos. We asked the following: (a) What issues, problems, and potential solutions consumed the discussion most; (b) how did the group's attention to various themes and interactive stances evolve and what does that show about the analyses and realizations of improvements that did and did not occur; (c) how were people's interactions, degree of centrality, and content shaped by roles; and (d) how did these dynamics affect the group in realizing improvements?

We went through multiple iterations when constructing our codebook. We initially framed our categorizing and selection of variables around the prevalent concepts in the cognitive map. We then reread the transcript and refined themes and presentation modes for coding. We used standard interactive and affective content analysis categories to record teammates' dispositions. We progressively refined our variables and their descriptions into the final codebook, excerpted in the Appendix. In addition, we recorded metadata—the message a paragraph belonged to, its author, recipient, its author and recipient's roles and organizations, and timestamp (day and time). We used paragraphs as the unit of analysis. Two raters coded 296 paragraphs and identified in each the presence or absence of each variable, resulting in dichotomous scoring. Out of a sample of 4,800 scoring decisions, coders agreed on average 97% of the time. The extent to which coders agreed on each variable is detailed in the Appendix. Also included in the Appendix is the extent of agreement between the coders about the absence or presence of each variable in the transcript. Inter-rater reliability is 0.68 based on the kappa statistic, which is used to measure agreement beyond agreement by chance for binary ratings (Cohen, 1960). We tested for an association between the coders, which is statistically significant.[1]

We analyzed the frequency and occurrence of various themes and relationships between themes and other variables and metadata to identify prevalent content in this postmortem e-mail disucssion. We examined the extent to which various content (or its absence) was associated with specific roles and the effects of roles on the group's problem-solving outcomes. We also identified social networks among teammates; the themes, interactive and affective stances that joined people; and the characteristics of "hub" or "end-point" writers. We identified patterns in exchanges by day, role, individual problem solver, and subproblem. These analyses, however, did not show the effects of teammates' rhetorical choices on problem-solving outcomes. Therefore, we turned to dramatistic analysis.

Dramatistic Analysis

As a rhetorical methodology, dramatistic analysis presumes that textual exchanges make certain realities present to participants and leave others out. People exchanging these texts choose some emphases and omit others to get readers to go along with them. In communicating emphases, Burke (1969) noted, discourse is action. It expresses the view the writer wants others to believe about some act being undertaken by some agent in a specific setting or context through some means for a purpose. These dramatic elements—act, actor, scene, agency, and purpose—are very much present in complex problem-solving debates.

Writers such as our e-mail participants evoke emphasis by combining dramatic elements into pairs or ratios (e.g., actor–action, actor–scene, etc.) and selecting one pair rather than another to tacitly highlight a certain point of view that they value. For example, a classic use of a scene–action ratio occurs in film noir when a dark, shadowy setting portends danger and violent actions.

---

[1]$p < 0.0001$, chi-square test of no association; statistical analyses were run using SAS 9.1.

We addressed the following issues using dramatistic analysis of e-mails: (a) patterns in emphases that participants cumulatively constructed through ratios for different subproblems and their effects, (b) tensions in motives revealed by these emphases, (c) connections between patterns in emphasis and conventions for coordinating cross-domain work in decentralized environments, and (d) the effects of discursive emphasis on the realization of improvements.

To code texts for emphasis we used Burke's (1969) definitions of dramatic elements and ratios. First, we identified the predominating ratio that a writer used to frame his point of view for each paragraph of his message, and then we abstracted the emphasis for the message as a whole. We used the message as the unit of analysis—as opposed to paragraphs in content analysis—because we needed to account for the effects of the context of a message, that is, how preceding e-mails and anticipation of subsequent ones affected a writer's choice of emphasis.

For our analysis, we grouped exchanges according to the main categories of thought and action targeted by participants' emphases. Five categories surfaced that strongly resembled the categories from the earlier methods. They are detailed in the Results section under the heading *Dramatistic Analysis*.

We traced the flows of emphasis in threaded messages and structured these threads by category of issue. We identified subtle shifts in emphasis and issue between senders and responders and assessed the effects these shifts had on social integration and the realization of technical goals. Last, we analyzed the extent to which social convergence and the realization of technical improvements were in tension.

Data and Participants

Our transcript of 102 e-mail messages and 296 paragraphs comprised exchanges among 30 network specialists over 13 days, the first day of which was real-time troubleshooting involving 11 specialists and 25 e-mails. We examined I-2 historical documents as well—meeting minutes, planning documents, reports—to identify and track persistent problems, prior solutions, and improvements.

Participants in this case represent 5 professional roles and 19 organizations (4 types) across the country. Of the 30 participants, 37% were network operators, 23% were network engineers, 23% were network researchers, 10% were application specialists, and 6% were scientific researchers (end users).

Participants all belonged to the I-2 e2e performance e-mail list and decided on their own to take part in the impromptu problem solving. They were all men, and their degree of familiarity with one another varied. They may have been together on other problem-solving teams previously, but ad hoc postmortem team dynamics differed for each case due to the variations in team composition.

RESULTS

Cognitive Mapping

Findings from modeling the e-mail exchanges as a cognitive map show the exchanges were structured into two streams of argument, one for the problem of insufficient knowledge management and one for the problem of systemic packet loss. In each argument, numerous causes and solutions were presented and related to nine expressed utilities, such as low costs, reduced packet loss, and in-synch network components.

Arguments related to the knowledge problem of incomplete information were linear and straightforward, with causes and solutions largely residing outside the inner workings of the network. After this e-mail exchange, technologists realized some of the improvements they discussed for this problem, such as adding new information to the Weathermap. However, they did not realize improvements that required sticky organizational prerequisites, such as providing dedicated resources or incentives for creating and populating knowledge bases, despite explicitly discussing the need for these prerequisites (see Figure 1a). Delving into these organizational issues to improve knowledge management would have been more difficult than undertaking such extra-infrastructure fixes as improving the Weathermap or other documentation.

In arguments for the second subproblem—causes of systemic packet loss— chains of reasoning were more complex and often involved contingent reasoning, relationships between multiple factors, trade-offs, and debates over goals (See Figure 1b). Unlike the problem of incomplete information, the causes of systemic packet loss were entrenched in the complicated inner workings of the system, involving incompatibilities between network components, applications, and protocols. Arguments and reasoning that participants advanced about the causes and resolution of this problem depended on the goals they set for network performance, and these goals differed. Some teammates were strong proponents of zero packet loss as the prime goal, and they argued in favor of better network design. Other teammates posited acceptable packet loss as the goal, which led them to claim that applications, networks, and transport protocols must all be redesigned. Members of this group disagreed among themselves about which of these design orientations had top priority and who was responsible. For the overall problem of systemic packet loss, the team did not realize improvements or reach consensus about the form and content the design solutions should take.

This lack of resolution was not new. Historical I-2 documents showed that debates about systemic packet loss and redesign had been going on for at least two years without much progress. One document, for example, called for developing common agreement on what constitutes "OK network service" and bridging the

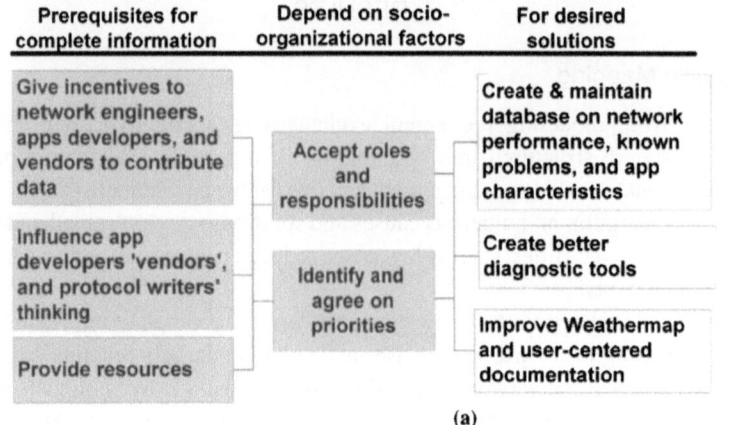

(a)

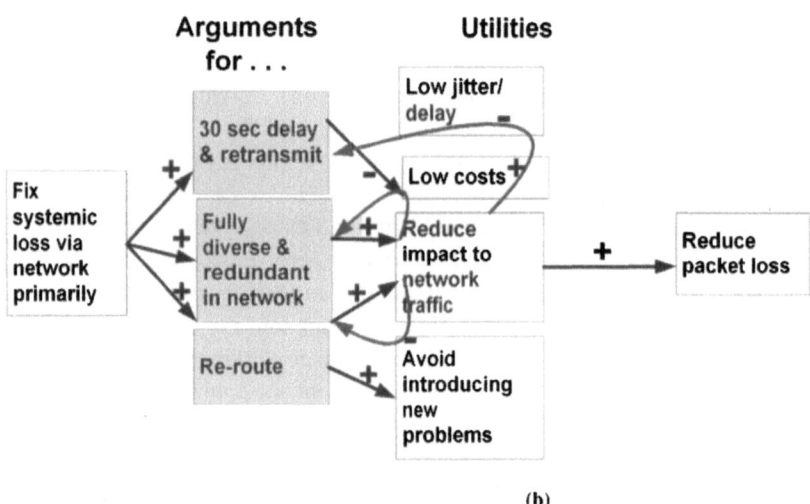

(b)

FIGURE 1   Part of the cognitive map. (a) Linked ideas about causes of incomplete information, framed as factors required for making information complete, including organizational factors that teammates discussed but did not develop. (b) Arguments about systemic packet loss, framed as factors for fixing it linked to utilities. Curved arrows show a gain in one utility (+) but loss in another (−).

gap between the network mentality and the user–application mentality, two points reiterated in the arguments represented in this case's cognitive mapping. The high level of abstraction that I-2 specialists used to discuss this problem may have impeded them from realizing concrete improvements. As the cognitive map shows, I-2 problem solvers did not mix abstractions with concrete details to discuss such in-depth, touchy subjects as priorities, responsibility, or criteria for deciding trade-offs; and, as research has shown, for problem solving this mixing is crucial.

Content Analysis

Results of the content analysis show that several themes bound the group together, but the top three themes made up 68% of all paragraphs and involved more participation from collaborators than all other themes. These themes were as follows: systemic packet loss due to network causes (23%), systemic packet loss due to application causes (22%), and insufficient e2e diagnostic tests for troubleshooters (23%; see the Appendix for more detail).

Socially, network engineers and operators contributed most heavily to all themes, including the top three during the 13-day problem solving, and together they accounted for 67% of the exchanged paragraphs. More network operators participated in the exchanges than did network engineers, but they contributed 60 fewer paragraphs in almost the same quantity of e-mails. Network researchers, equal in number to network engineers, were a distant third in contributions. This distribution of problem-solving messages gave the team a decidedly network cast; application specialists and end users were clear minorities.

Despite the network orientation, most topics received an even distribution of messages from all roles. Yet roles had distinct profiles, and some dominated in certain topics and in expressions of agreement and opposition, as follows:

- Network engineers talked more than others about troubleshooters' diagnostic tests, even though network operators as the first line of defense in troubleshooting had an equal stake in them. But defining requirements and developing advanced e2e troubleshooting tools and tests were the engineers' specialty, and as gurus they dominated this topic. Many of the network engineers' messages examined new tools as the best hope for tackling incomplete information and system incompatibilities. They contributed 21% of the disagreements and 35% of the agreements articulated in the e-mail transcript.
- Network operators talked most about packet loss due to congestion, suggesting their comfort with knowable problems and concrete and familiar testing methods and tools. Attesting to their hands-on orientation, network operators called for action three times more than any other group. They also did the most writing about accessing and disseminating information from hardware vendors and educating users—the two outside groups who they knew best and who they believed needed to share and receive knowledge better. They contributed 43% of the disagreements and 35% of the agreements.
- Network researchers dominated discussions about trade-offs, a complex topic that involves investigating the conditions under which it is acceptable to sacrifice one good measure of performance for another (e.g. speed, latency,

packet loss, and jitter). In their messages, researchers sought information and ideas from teammates more than did any of the other participants and called for action less. They contributed 14% of all disagreements and 4% of agreements articulated in the e-mail transcript. More than any other participants, they often sent messages that did not address any theme. Their content patterns reveal a profile of argumentative inquiry, a mission to gather and analyze data, and a distancing from actions that might have grown out of this discussion.

- Application specialists, in keeping with their interests, devoted more of their paragraphs to application-design issues relevant to systemic packet loss than to any other theme. But, reflecting their minority position, the application specialists' combined contributions barely equaled that of one network engineer who apparently felt as strongly as they did about systemic packet loss. Application developers articulated less disagreement (7%) than did most of those in other roles, despite their concentrated attention on contested systemic-packet-loss issues. Support was their most prominent interactive stance and contributed 17% of the agreements expressed in the e-mails. They struck agreeable stances even for touchy subjects.

- End users dominated none of the themes, demonstrated no priority themes, never discussed collaborations, and had no clear patterns of contribution other than being willing to voice disagreement in this forum. End users contributed 7% of the disagreements and 9% of the agreements articulated in the e-mail transcript.

In cumulative discussions, ideas were negotiated, advanced, and aborted based on who talked to whom in exchanges of consequence. Everyone's exchanges and responses freely crossed role boundaries, and the foundation for effective consensus building was established. People who fielded the greatest number of responses were the hubs, eliciting at least three replies to one of their messages, and most of the hubs were network operators. Their messages typically contained multiple themes. Consensus building seemed to thrive when a sender provided a lot of material that required simultaneous consideration.

At the opposite end of the spectrum of influence were end points, messages that received no attention from teammates. Of all messages, 50% were end points. Many of the end points (35%) were requests for responses issued by researchers about trade-offs. It is hard to know why end points occurred, but a failure to respond to the theme of trade-offs closed opportunities for the group to discuss the priorities and criteria required to realize improvements under different conditions.

The research literature has established that disagreement is a prerequisite for consensus building. However, our findings about the group's interactive stances show that little conflict occurred. Only 4% of the paragraphs expressed opposition to teammates' messages. Moreover, disagreement was absent until the fifth day, when systemic packet loss began to consume the conversation, and underlying causes, solutions, and goals became ambiguous. Agreement, by contrast, began and continued from the first day on; it addressed a greater diversity of topics and existed in twice as many paragraphs.

The group rarely addressed conflict management strategies. Nor did they often call for explicit action or collaborative efforts. When calls for action did occur, they came mostly from network operators requesting solutions to the problem of keeping everyone better informed. Commitments to collaborate often went unremarked. Interactive stances show that the group engaged much less in explicitly dealing with the social dynamics of their varying points of view than in discussing technical aspects of the problem. As the literature has shown, too little conflict can be as troublesome to consensus building as too much because a lack of conflict shuts off potentially useful and novel approaches to understanding and resolving complex problems.

Dramatistic Analysis

A rhetorical analysis of the dramatic emphasis in problem solvers' e-mails showed that participants highlighted various ratios (actor–action, action–scene, actor–agency, etc.) as they progressively discussed the following five broad issues: (a) commitments to and beliefs about various goals, (b) technical accountabilities in systemic packet loss, (c) human accountabilities in systemic packet loss or incomplete information, (d) consensus-generating moves highlighting actions or summaries, and (e) technical accountabilities in the problem of incomplete information.

As seen in the other textual analyses dealing with the problem of incomplete information, participants' rhetorical moves for emphasis were fairly uncomplicated and linear. Teammates emphasized three main issues—human and technical accountability and consensus-related actions or summaries. In Figure 2 each is indicated as a separate row of messages plotted against time. Typically, teammates progressively built on each others' emphases. For example, in Figure 2, steady, horizontal links between messages in the *Technical Issues, Incomplete Information* row, starting on the first Friday, represent a sequence of senders and responders who maintained a consistent emphasis on this issue. Some of the messages about incomplete information shifted emphasis—mostly to human issues or consensus-related acts or summaries—and vertical lines indicate this shift, linking a message in the Technical Issues row to a response in another row. When emphasis shifted, subsequent responders typically sustained a sender's emphasis for at least a few messages before turning the conversation back to technical issues. Only two messages, both on the first Tuesday, attempted to tie together the problems of incomplete information and systemic packet loss, but with no follow-up from others.

By contrast, problem solvers were nonlinear in the emphases they used to move discussions along for the problem of system packet loss. Figure 3 shows several

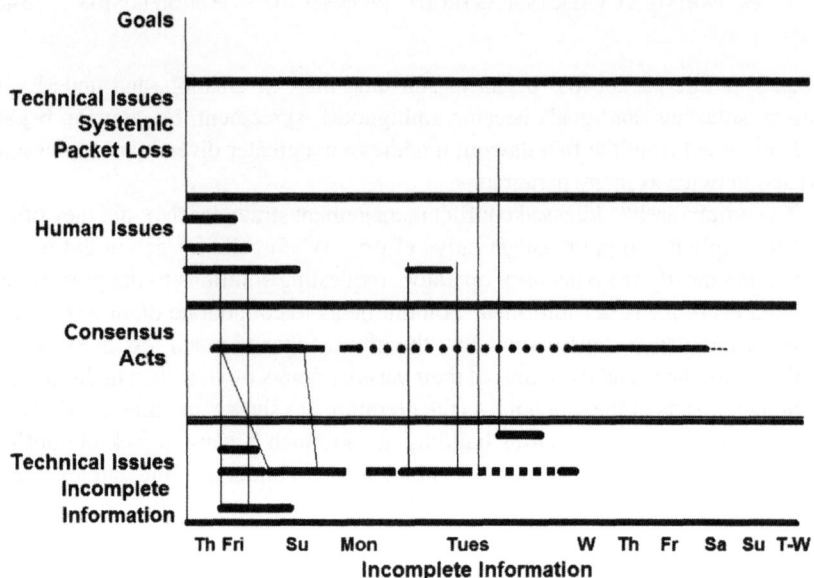

FIGURE 2 Issues discussed about incomplete information. Dotted horizontal lines connect a responder to sender when the response comes after a lapse in time. Vertical lines show a response that leapt to another issue or emphasis.

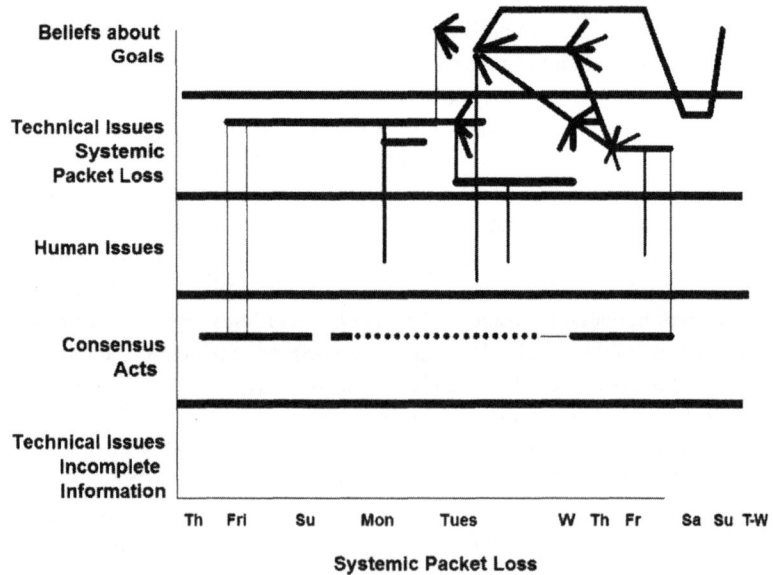

FIGURE 3 Issues discussed about systemic packet loss. Dotted horizontal lines connect a responder to sender when the response comes after a lapse in time. Vertical lines show a response that leapt to another issue.

344

star formations in which lone points of view emanate from a motivating source message and go unanswered. Structurally, team members made frequent, unsystematic leaps across issues to turn emphases toward their own diverse biases, many of which, as the vertical links indicate, were never picked up and carried on by others. Rhetorical analysis of these irregular shifts in emphasis revealed several underlying motivations. These included a desire to mask disagreements, maintain strategic ambiguity for decentralized coordination, and avoid direct discussions about domain-specialists' responsibilities. We now explore our findings about these motivations.

Masking disagreement.   In inquiring into systemic packet loss, problem solvers debated whether accountability lay with networks, applications, or protocols, and occasionally these exchanges came close to blaming people or dictating actions. Teammates avoided this, however, by finessing their emphases rhetorically.

In one exchange, a network engineer commented that applications were not designed to be sensitive to network conditions and, therefore, were the prime cause of systemic packet loss. He emphasized that applications as actors precipitated adverse actions (systemic packet loss). In response, an application specialist seemingly agreed but in the same breath shifted the emphasis to the network environment as the prime actor influencing network performance. Shifting to a scene–action emphasis, he argued that faulty application behaviors (still the actions) were, in fact, products of their network environments (scenes). He claimed that an application might cause problems in one network but be a "model citizen" in another. At this point, a third participant joined in and shifted the emphasis back to the application as actor, but now with a different cast. This writer agreed with the application developer that various applications worked more and less well, depending on the network, and he detailed which network conditions required what application behaviors. Given the availability of this technical knowledge, this writer argued, applications (actors) should be able to adjust their behaviors (agency) to network conditions. In this actor–agency emphasis, applications must be self-regulating.

Three positions or emphases were in play: applications as culprits, applications as victims, and applications as self-managing agents (through built-in intelligence). These different biases, however, went uncontested. On the contrary, participants made explicit moves to agree with each other. For example, the network engineer who came close to blaming the application as culprit asked the others to critique his position to be sure "we're in sync." One person responded immediately, "Yes, we are in synch." The other person, the application specialist, waited a day to respond and then reasserted his point of view without addressing the question of being in synch.

The application specialist may have been silent about being in agreement or disagreement with the other participant's emphasis on the application-as-culprit, but by virtue of his continued communications, he tacitly conveyed he was at least in synch with the conventions taking shape for talking about this tense subject. These conventions involved subtly shifting emphasis but presuming agreement despite Rashomon-style disagreements. These conventions enabled teammates to move analyses of complex system incompatibilities forward. But they also resulted in people shifting emphasis before any one rendition of an explanatory story could develop or could be explored thoroughly, as is required for a comprehensive complex analysis.

Maintaining strategic ambiguity.   As Figure 3 shows, midway through the communications about systemic packet loss, participants introduced a stream of exchanges to address goals. This shift in attention to goals seems to have been motivated by the group's desire to maintain strategic ambiguity in the face of potentially heated debates. One writer, for example, explicitly stated, "Talking about how to make applications more robust ... is a fine thing to do but it also seems to be taking the focus off of something we really need to do to achieve good end to end performance, that is, run our networks well."

Ironically but not surprisingly, the rhetorical shift to a shared goal of running the network well led to disagreements. As shown in earlier findings, problem solvers debated whether the goal should be zero or acceptable packet loss, or as one network operator put it, the pristine or the real. Nonetheless, contentions over goals were far less serious than the large rifts would have been had the team analyzed in depth accountabilities for system incompatibilities. Maintaining their emphasis on high-level purposes safeguarded them from serious role-based conflicts. Unfortunately, this emphasis also prevented the detailed analysis and conflict required to negotiate design issues.

Human, domain-based responsibilities.   Just as the group sustained strategic ambiguity by shifting its emphasis to goals, it also strategically avoided debates about the bounds of each functional area's responsibilities in terms of who owned what potential redesigns. In one spate of e-mails, 7 of 10 sequential messages broached the issue of domain-based responsibility by highlighting human actors and actions as the emphatic pair; for example, "We [network engineers as actors] are forced to engineer and operate [actions] our networks for the lowest common denominator apps." This writer was describing and rationalizing how a group of domain specialists approached their design responsibilities. As the group continued delving into distinct domains' responsibilities, an application specialist offered an example filled with technical details. The team gravitated toward the (safer) technical details and began reorienting the conversation from human actions to technical behaviors and detailed system interactions. The technology and its inner workings became surrogates for humans but not due to any leanings toward technical determinism in the team. Rather, technology surrogates for humans functionally served the same purpose as talking about goals at a high level. In both cases, collaborators turned to the familiar to assume similarities in identity and interests rather than disparities, assuring decentralized coordination and social integration.

DISCUSSION

Overall, I-2 teammates were quickly and straightforwardly able to tackle the most obvious, surface problem of incomplete information. However, they skirted many of the thorny organizational requirements needed to realize solutions, such as resources, incentives, training, and creating and using better knowledge bases and documentation. For certain improvements, like the Weathermap additions, the group used a decentralized first-come, first-served strategy to implement solutions that had little organizational overhead. Teammates treated the ambiguous problem of systemic packet loss and the redesigns it would require differently. This problem was intellectually complicated due to technical and organizational interdependencies and contested subgoals. First-come, first-served strategies as a means for implementing improvements were not likely to work.

Realizing improvements hinged on the limitations imposed by the social dynamics of collaborative problem solving as much as it did on technical know-how and feasibility. Results from cognitive mapping suggest that distance-collaboration approaches to problem solving among diverse specialists combined with these individuals' adherence to conventional argumentation structures may have kept them from exploring in detail many issues and interconnections that they introduced on a general and abstract level. Conceivably, conventional causal-effect structures may have encouraged them to compartmentalize diverse factors, which becomes counterproductive for complex problems in complex systems when problems are caused by incompatibilities across technical and human domains. Results from content analysis add to these insights. They reveal that who people talked to and what they talked about subtly constrained the technical and design issues that teammates allowed each other to bring to the table. For example, without much vocal opposition from application specialists in their area of expertise—application design—debates about systemic incompatibilities and solutions for them remained conceptual and hypothetical rather than pragmatic. At this level of abstraction, decisive conclusions are less likely.

Disagreement and conflict-management strategies play important roles in social integration and in problem solving, but they function differently in each. In this team, overt disagreement rarely occurred, seemingly a strategy for social integration and convergence. Expressed disagreement often received no response, and people defused conflict by hitching opposition and consequential comments to a multiplicity of ideas, allowing teammates to respond to the less volatile issues. However, these moves toward social integration constrained the technical ideas and improvements that teammates talked about. In addition to premature closure of conflict, the team may not have realized improvements because inquires did not adequately span and integrate every level of abstraction—high-level goals, low-level inner working details, and pragmatic-level dependencies and causal and conditional interactions. In addition, communications rarely made explicit an awareness of group processes or problem-solving progress.

It is tempting, based on this assessment, to speculate how participants might have shaped social dynamics to better realize improvements. For example, what would have happened if application developers voiced more disagreement or pushed to dominate the theme closer to their self-interest? What if end users were less peripheral? What if more participants than researchers took up the pragmatic-level issue of acceptable trade-offs? What would have occurred if people revisited action items and if responders asked who should be responsible for what actions? What if some people overtly assumed the roles of facilitator or awareness monitor to make individual positions more explicit?

Unfortunately, these questions and their implicit recommendations do not adequately consider the complex array of problems that the collaborators rhetorically strived to manage. They constructed and stabilized conventions in language acts for discussing potentially charged issues and ambiguities. They enacted emphases that would evoke a feeling of consensus without calling domain-specific assumptions into question and without threatening the delicate balance of autonomy and interdependence that connected decentralized functional areas. From this perspective, the I-2 problem solving negotiated a social and intellectual minefield.

Through subtle shifts in emphasis, teammates were able to simultaneously agree and disagree. They strategically and ambiguously focused on high-level goals to establish a sense of common ground while still carrying out a complex analysis, albeit for the more benign issue of goal selection. They subtly personified technologies and averted dictating responsibilities to others without sidestepping their inquiry into complex interactions that caused system incompatibilities and systemic packet loss.

However, to analyze complex root causes of systemic packet loss, to find design solutions, and to figure out how to turn them into implemented improvements, collaborators needed different sociocognitive processes. As Weick (1995) argued, "The problem in ambiguity is not that the world is imperfectly understood and that more information will remedy that. The problem is that information may not resolve misunderstandings" (p. 92). Intrinsically, social integration and cognitive complexity merge when resolving misunderstandings but not in the communicative ways enacted by the collaborators in this case. Rather, problem solvers need to scrutinize unclear means–end relationships in goals, technologies, and organizational dependencies from various and often contentious perspectives. They need networked views of multifaceted causes for e2e, cross-domain network problems. They need to wrestle with inferences drawn from different role-based orientations and negotiate their conflicting interpretations. They often need to arrive at novel options, which require some degree of real cognitive dissonance and an immersion in levels of functional analysis that the group skirted in favor of either too high or too low a level. They need to realize that eventually they must make a judgment call when no one right answer is available.

But in their problem solving, teammates were torn. Shifts in dramatistic emphasis seemed to enable collaborators to veer away from conversations that might uncover deep disparities while still working on root causes of systemic packet loss. Collaborators achieved swift trust, agreement, and a cooperative spirit, but these came at the expense of paying too little attention to ambiguities that had to be resolved—for example, ambiguities about priorities, trade-offs, and means–end relationships. True consensus about the best ways to resolve system incompatibilities by design would have required examining these issues As a result, participants ultimately moved away from their initial goal of arriving at improvements that would prevent the same transmission failure from occurring in the future and instead turned brainstorming and sharing intellectually stimulating arguments and engaging case stories into ends in themselves. The group detached this set of inquiry processes from the other problem-solving processes required for coming to conclusions and deciding next steps. Ironically, problem solvers in this forum may have been able to sustain the high degree of complex exploration and participation because they sensed the group's purpose had imperceptibly changed to a less threatening, albeit ultimately less useful goal.

CONCLUSIONS

Complex systemic-packet-loss improvements did not get realized in this collaboration, and they remain elusive to this day. Realistically, ad hoc remote teams of diverse specialists are not going to decide on solutions to systemic problems in 70 or so e-mails. But given the intensity, intelligence, and time invested in these discussions, our research team expected to see more progress than occurred in making designs for system incompatibilities a top priority.

This study shows that understanding the social dynamics in collaborative problem solving for complex technical faults is integral to the technical improvements that a group realizes. Technologists need to recognize the workings and effects of this intertwining. As our transcript analysis shows, when technologists do not have this understanding, they may choose collaborative interactions and communications that counteract and constrain their own good intentions for achieving technical improvements without recognizing it.

In decentralized, distributed contexts, cross-functional collaborators have to reconcile the social dynamics required for coordination and social integration with those required for exploring complex, ambiguous problems and making consensus-based decisions about technical design. To better realize improvements, we suggest three perspectives that run counter to common assumptions related to technology diffusion:

1. Improved tools cannot fix everything. At the root of many e2e distributed-computing problems is system design, and tackling its unclear boundaries for improvements requires that technologists treat the need to develop appropriate group processes as seriously as they treat the development of advanced tools.

2. System design improvements cannot be put off endlessly in favor of solving more straightforward problems. Doing so just moves problems around. Without system-design improvements, systemic problems are sure to recur in other parts of the system, and this vulnerability obstructs diffusion readiness.

3. Successful problem solving for diffusion-based improvements must involve two essential but different types of flexibility: the flexibility required for social cohesion and the flexibility required for consensus-based problem solving and decision making. Teams need to explicitly increase their awareness about how these types of flexibility differ, for example, levels of abstraction appropriate for each. With this awareness, collaborators may better develop overt and tacit efforts to accept yet balance these tensions to move toward decisions (rather than, for example, recasting goals so that brainstorming becomes an end in itself).

Our case study shows that bringing these perspectives into virtual collaborative problem solving is a communication act, and teams need greater awareness of the specific factors within this act that affect technical goals. Communication experts, either as consultants or core members of technology teams, can facilitate technologists' success in gaining and articulating this awareness. Moreover, researchers in technical communication and other areas in human–computer interaction can inquire further into the distinct sociodiscursive dynamics required for effectively analyzing and solving complex problems to ready innovations for diffusion. Our study demonstrates that we need more studies specifically focused on complex problem solving, especially for ambiguous problems, because effective common-grounding processes and levels of analysis are different in kind not just degree from straightforward problem solving.

ACKNOWLEDGMENTS

We wish to acknowledge and thank Lisa Levy, Rod Johnson, and Norbert Elliott for their excellent guidance, insights, and suggestions, and the people at I-2 for their support. We also thank Jing Deng for her critical contributions.

REFERENCES

Axelrod, R. (Ed.). (1976). *Structure of decision*. Princeton, NJ: Princeton University Press.

Barkhi, R., Jacob, V., & Pirkul, H. (2004). The influence of communication mode and incentive structure on GDSS process and outcomes. *Decision Support Systems, 37,* 287–305.

Birnholz, J., Finholt, T., Horn, D., & Bae, S. J. (2003). Grounding needs: Achieving common ground via lightweight chat in large, distributed, ad-hoc groups. *Proceedings of the Conference on Human Factors in Computing Systems* (pp. 21–30). New York: Association Computing Machinery.

Burke, K. (1969). *A grammar of motives*. Berkeley: University of California Press.

Cohen, J. (1960). A coefficient of agreement for nominal scales. *Educational and Psychological Measurement, 20,* 37–47.

Coppola, N., Hiltz, R. S., & Rotter, N. (2004). Building trust in virtual teams. *IEEE Transactions on Professional Communication, 47,* 95–104.

Ham, D., & Yoon, W. C. (2001). The effects of presenting functionally abstracted information in fault diagnosis tasks. *Reliability Engineering and System Safety, 73,* 103–119.

Herbsleb, J., Mockus, A., Finholt, T. A., & Grinter, R. W. (2001). An empirical study of global software development: Distance and speed. *23rd International Conference on Software Engineering* (pp. 81–90). Washington, DC: IEEE Computer Society.

Hollinghead, A. B. (1998). Retrieval processes in transactive memory processes. *Journal of Personality and Social Psychology, 74,* 659–671.

Jarvenpaa, S., & Leidner, D. E. (1999). Communication and trust in global virtual teams. *Organizational Science, 10,* 799–815.

Kramer, A., Fussell, S., & Setlock, L. (2004). Text analysis as a tool for analyzing conversation in on-line support groups. *Proceedings of Conference on Human Factors in Computing Systems, Late Breaking Papers* (pp. 1485–1488). New York: Association of Computing Machinery.

Malone, T., & Crowston, K. (1994). The interdisciplinary study of coordination. *Association of Computing Machinery Computing Surveys, 26,* 87–119.

Meyerson, D., Weick, K. E., & Kramer, R. M. (1996). Swift trust and temporary team. In M. Kramer & T. R. Tyler (Eds.), *Trust in organizations* (pp. 166–195). Thousand Oaks, CA: Sage.

O'Rorke, P., & Henrion, M. (1996). *Acquiring knowledge relevant to tasks involving decision making (Tech. Rep. SS-96-0)*. Retrieved December 26, 2005, from http://overcite.lcs.mit.edu/rorke96acquiring.html

Panteli, N., & Sockalingam, S. (2005). Trust and conflict within virtual inter-organizational alliances: A framework for facilitating knowledge sharing. *Decision Support Systems, 39,* 599–617.

Postmes, T, Spears, R., & Lea, M. (2000). The formation of group norms in computer-mediated communication. *Human Communication Research, 26,* 341–371.

Schaafstal, A., Schraagen, J. M., & van Berlo, M. (2000). Cognitive task analysis and innovation of training: The case of structured troubleshooting. *Human Factors, 42,* 75–86.

Stålhane, T., Dingsøyr, T., Hanssen, G. K., & Moe, N. B. (2003). Post mortem—An assessment of two approaches. In R. Conradi & A. Wang (Eds.), *Lecture notes in computer science: Vol. 2765. Empirical methods and studies in software engineering* (pp. 129–141). Berlin, Germany: Springer-Verlag.

Strang, D., & Meyer, J. (1994). Institutional conditions for diffusion. In W. R. Scott et al. (Eds.), *Institutional environments and organizations: Structural complexity and individualism* (pp. 100–112). Thousand Oaks, CA: Sage.

Tidwell, L., & Walther, J. (2002). Computer-mediated communication effects on disctime. *Human Communication Research, 28,* 317–348.

Walther, J. (2004). Language and communication technology: Introduction to the special issue. *Journal of Language and Social Psychology, 23,* 384–396.

Weick, K. (1995). *Sensemaking in organizations*. Newbury Park, CA: Sage.

Wellman, M. (1994). Inference in cognitive maps. *Mathematics and Computers in Simulation, 36,* 137–148.

Wrightson, M. (1976). Documentary coding method. In R. Axelrod (Ed.), *Structure of decision.* Princeton, NJ: Princeton University Press.

Barbara Mirel is a research faculty member in the School of Information at the University of Michigan and teaches courses in information visualization and digital discourse. She is the author of *Interaction Design for Complex Problem Solving: Designing Useful and Usable Software* (Elsevier/Morgan Kaufmann, 2003) and is the co-editor (with Rachel Spilka) of *Reshaping Technical Communication,* which was awarded the 2003 Best Collection of Essays Award for the National Council of the Teachers of English Excellence in Technical and Scientific Writing. She has published numerous articles and book chapters, and in 2000 she received the ACM-SIGDOC Rigo award for lifetime achievement.

Nicholas Johnson holds a Master of Science degree in Information Science, focusing on human–computer interaction, from the University of Michigan. He also received a Bachelor of Arts degree, with high distinction, in Cognitive Science from the University of Virginia. While at the University of Virginia, his research into realistic human walking behavior received the Radar Award for Undergraduate Research. He now works at the U.S. Department of Agriculture Economic Research Service, where he performs usability engineering and process improvement.

Codebook and Scoring Outcomes

	Coder 1, Absence of Variable (N = 75 Paragraphs)	Coder 2, Presence of Variable (N = 75 Paragraphs)	Coder Disagreement
Variable 1: Level of Detail/Presentation Mode			
Story	71	2	2
Link/Citation/Raw data	68	5	2
Variable 2: Interactional Analysis			
Offering debatable position	27	33	15
Offering opinion	70	2	3
Seeking information	73	2	0
Seeking idea or position	72	1	2
Supporting	66	5	4
Opposing idea/information	65	3	7
Variable 3: Affective Stance			
Apologetic	73	2	0
Humorous	70	0	5
Clout of personal credibility	68	1	6
Variable 4: Theme			
Systemic packet loss—application causes	49	9	17
Diagnostic tests for troubleshooting	59	14	2
Systemic packet loss—network causes	50	9	16
Solving the problem event	75	0	0
Reference to the problem event	71	2	2
Quality of service issues	74	0	1
Trade-offs related to packet loss	65	4	6
Packet loss and congestion	75	0	0
Smarter application developers	71	1	3
Accessing/disseminating application information	75	0	0
End-user diagnostic tests and tools	59	14	2
Collaboration	70	0	5
Information contributions from NOCs	75	0	0
Accessing known problem information	72	0	3
Educating end users	75	0	0
Information contributions from application developers	75	0	0
Information contributions from hardware vendors	75	0	0
Accessing/disseminating network information	75	0	0
Collaboration	70	0	5
Variable 5: Action Item			
Call for group or individual action	70	5	0

Note. Variables and coder agreement and disagreement about variables in a sample of 75 paragraphs, 25%, of the e-mails (ID Nos. 18–21, 29–33, 72–81, 93). Because of the sampling, some variables present in greater proportion in the text (e.g., "Solving the problem event" under *Theme*) may be underrepresented here. NOCs = network operations.

TECHNICAL COMMUNICATION QUARTERLY, *15*(3), 355–382

A Hybrid Analytical Framework to Guide Studies of Innovative IT Adoption by Work Groups

David Dayton
Towson University

This article presents a framework for analyzing innovative information technology adoption by organizational work groups. Concepts from three distinct theories (adoption and diffusion theory, cultural–historical activity theory, and the social construction of technology) are modified and integrated to form a hybrid, layered framework, which is then applied to a specific case to demonstrate the advantages for guiding research and analysis. The illustrative case presents the experience of a small work group in a high-technology company that implemented single-source content management.

The primary purpose of this article is to present an analytical framework to guide qualitative studies of work groups as they collectively learn, analyze, adopt, and redefine a new information technology (IT) tool or system. The framework I propose blends the structural and heuristic concepts of Rogers's (1962, 2003) adoption and diffusion theory (ADT), cultural–historical activity theory (CHAT; Engeström, 1999a; Leont'ev, 1978), and the social construction of technology (SCOT; Pinch & Bijker, 1984). To illustrate the analytical utility and explanatory power of this framework, I apply it to the case of a small technical communication work group that abandoned Adobe FrameMaker and Quadralay WebWorks Publisher in favor of AuthorIT, software that enables the group to publish print manuals, create a dynamic hypertext knowledge base, and develop online help—all from a single database of structured content. AuthorIT and some of its competitors refer to this type of innovative IT tool as a single-source content management (SSCM) system.

To set the stage for the retelling and analytical re-visioning of this case, I first situate the study of IT adoption and use within the literature of our field and show its connections to research in other IT-related disciplines. In the second section, I argue the particular importance for our field of studying the adoption and use of

SSCM by technical communication work groups. I then explicate the analytical framework I am proposing. In the fourth section, I use the framework to structure a narrative and analytical discussion about the illustrative case of SSCM. My ultimate goal is to demonstrate the usefulness of this hybrid, layered framework for studying the adoption and use of innovative IT by organizational work groups.

FROM DISCIPLINARY ROOTS
TO CROSS-DISCIPLINARY BRANCHES

In *Rhetoric, Innovation, and Technology: Case Studies of Technical Communication in Technology Transfers*, Doheny-Farina (1992) examined technology transfer through the lens of social-constructionist rhetoric. He demonstrated that the *transfer* aspect of the phenomenon is a misnomer, arguing persuasively that "technology transfer is a rhetorical dynamic" (p. 4):

> Every aspect of technology transfers must be negotiated, constructed, and reconstructed in the minds of the participants. There is no clearly objective fact or physical entity that proceeds uninterpreted from the lab to the market. The entire process is one of interpretation, negotiation, and adjustment. Moreover, it engenders a reciprocal shaping as it develops; the innovators, the innovation, and the users of the innovation are all changed through the process. Additionally, what any one participant perceives as the technology or the innovation may be different from what others perceive. (p. 6)

Doheny-Farina (1992) opened a new research space for technical communication research by framing his case studies of business and technical writing in the overlapping area of a Venn diagram whose distinct circles are rhetoric, innovation, and technology. Over the past decade, our field has been busy and productive in two related work zones that often occupy that research space. Our research and theory have broadly examined and deeply analyzed the interplay of sociotechnical and rhetorical dynamics at work in

1. The social construction of information and communication technologies, the bidirectional effects shaping them, and the writing activities and subjectivities they mediate (e.g., Duin & Hansen, 1996; Haas, 1996; Sullivan & Dautermann, 1996).
2. The knowledge-making activities in professional disciplines and within and among business, governmental, and nonprofit organizations (e.g., Bazerman & Russell, 2002; Freedman & Medway, 1994).

Studies in our field that fit squarely within one or partially in both of the two categories just mentioned are numerous. Most of those focusing on a new technology are impact studies that describe and analyze the changes to discursive work activities resulting from use of a new communication technology such as word processing, e-mail, online documentation, the World Wide Web, and, most recently, SSCM. Within this same general category, media-choice studies constitute a secondary stream of research, reflecting our interest in computer-mediated communication. A small number of studies falling in the first of the aforementioned categories focus specifically on the sociotechnical and rhetorical dynamics shaping the IT adoption process of groups or individuals.

In his multiple case-study report, Hansen (1996) observed and interviewed four knowledge workers in the same company about their use of the various communications technologies available to them, principally voice mail and e-mail. He learned that each person had a distinct pattern of using these tools, and all had sound pragmatic reasons for their choices. He concluded that "the use of different communications technologies reflects the preferences of different personalities" (p. 321), as well as job functions and different perceptions of the social context in which their interactions with others are embedded.

Kahn (2000) took a contrastive case-study approach grounded in structuration theory to examine the impacts of adopting digital technologies on the archives offices at two universities. In one case, the innovation-adoption process led to improved performance and status, whereas in the other it exacerbated the office's declining fortunes. Sociotechnical conditions were particularly positive and helpful for the successful office, but a more important factor in its success was a rhetorically savvy strategy for managing change, which contrasted sharply with the failed office's arhetorical, passive, and isolating interpretation of its new technology.

Downing (2004) interviewed and observed more than 50 workers at four branches of a call-center provider, focusing on technicians' use or nonuse of knowledge management software. He reported that a mix of sociotechnical and rhetorical factors contributed to widespread resistance to the new tools. For example, the tools undermined the socializing routines of technicians, who liked the occasional break and informal talk with colleagues prompted by difficult questions (p. 180). A de-skilling design forced technicians to conform their problem-solving and information-seeking routines to fit the program's decision tree, something that many of them were simply unwilling to do. Another factor working against use of the tools was the negative bias toward them among technicians who had tested the buggy beta versions. These early adopters were reluctant to use the final version of the software and discouraged others from using it (p. 179).

Rehling (1999) reported a case study of a technical publications group whose eagerness to adopt online documentation technologies led to adverse structural changes, creating a cautionary tale that ran counter to the unmitigated success story told by management. Rehling uncovered a counter-text whose theme was that management's infatuation with the latest and greatest technology resulted in documentation that was less usable for end users and in a loss of status and operational coherence for the documentation group.

First-person participant reports focusing on IT adoption and use form a subset in this research work zone. Beason (1996) reported on a two-year participant–observer study he conducted while employed as a technical writer at a computer company. In an analytical narrative rich in cultural–historical detail, he described how writing for publication on the Web changed the team's documentation-development process, encouraging a more collaborative approach while redefining their basic documentation genre to make it shorter.

Most recently in this vein, Kramer (2003) described his experiences in using a particular SSCM system developed in-house at IBM. His report highlighted the complexity of the system to illustrate how it changed the division of labor, requiring that writers learn challenging new technical and problem-solving skills. Kramer's account echoed Rehling's (1999) cautionary tale and reminded us that "every participant along the way constructs and reconstructs the innovation based upon his or her experience and worldview" (Doheny-Farina, 1992, p. 7).

In sum, our literature has shown that we have established a firm foothold in studying IT adoption and use. And it has demonstrated convincingly that we know how to open the black box of a specific context of activity and elucidate the mediation of collaborative work through discourse and technology. I see an opportunity here for us to raise our field's visibility and status by contributing to the knowledge construction going on—largely oblivious of our existence—at the intersection of business management, organization science, and information systems and technology.

Within that domain—information science, technology, and management (ISTM)—the dominant research paradigm is decidedly positivist; however, interpretive and critical paradigms constitute an alternative but highly influential approach (Fichman, 2000; Orlikowski & Baroudi, 1991; Walsham, 1995). Our field has natural affinities with the body of work produced by qualitative researchers in ISTM. To cite the most obvious example, Wanda J. Orlikowski of the Massachusetts Institute of Technology is a leading figure in the ISTM interpretive tradition and has conducted widely cited case-study research on the adoption and diffusion of IT innovations within organizations (see, e.g., Orlikowski, 1992; Orlikowski & Gash, 1994). Many of scholars in workplace communication have frequently cited genre studies in organizational communication conducted by Orlikowski and her collaborators (e.g., Orlikowski & Yates, 1994; Yates & Orlikowski, 1992).

The positivist paradigm in ISTM has become increasingly sophisticated at constructing surveys accounting for about two thirds of the variance between adopters and nonadopters of an innovation (Venkatesh, Morris, Davis, & Davis, 2003, p. 425); however, within ISTM it has been widely acknowledged for more than two decades that the precise nature of the social and cognitive factors that most influence any given innovation within a particular context is unique and likely to be complex (Blomberg, 1986; Venkatesh et al., 2003, p. 470). Recently, disciplines comprising ISTM have been exposed to a novel vision of organizational social psychology that supports a renewed emphasis on qualitative approaches to the study of innovative IT adoption and use.

Building on ideas central to the groundbreaking book in organizational psychology by Weick (1995), *Sensemaking in Organizations*, Swanson and Ramiller (2004) compared and contrasted mindfulness and mindlessness in organizations' adoption of IT innovations. In an article that won the 2004 best paper award from *Management Information Science Quarterly*, they wrote: "Mindfulness as the nuanced appreciation of context and ways to deal with it lies at the heart, we believe, of what it means to manage the unexpected in innovating with IT" (p. 256). Swanson and Ramiller's (1997) concept of mindfulness has roots in their earlier concept of any given innovation's "organizing vision," which they defined as "a construction in discourse" (p. 556), prominently citing Foucault (1972) and Porter (1990).

ISTM seems poised to take a rhetorical turn in analyzing innovative IT adoption and use. The time for us to look for cross-disciplinary research opportunities is ripe, and worthwhile sites for study are plentiful. For example, case studies of successful and failed IT innovation in hospitals have concluded that complex sociotechnical contexts can present obstinate difficulties for administrators trying to transfer new IT tools and processes (Lapointe & Rivard, 2005; Meyer & Goes, 1988). However, such studies have lacked in-depth analyses of the rhetorical clashes and mutually transforming interactions of groups' interpretive frames when innovation participants and stakeholders attempt to understand, refuse to use, or attempt to fit these innovations into their knowledge-creating activities.

STARTING IN OUR OWN BACKYARD: SSCM

My own interest in IT adoption and use has thus far centered on the technology-mediated work of technical communicators. In an extended study using multiple qualitative methods and a large-scale survey, I investigated the adoption and diffusion of electronic editing among practitioners (Dayton 2001, 2003, 2004a, 2004b). In analyzing the qualitative data from five sites where I interviewed and observed technical communicators, I was aided by two theoretical frameworks: CHAT, which is well known to our field, and Rogers's innovation ADT, which I had not found in our literature previously. In the culminating report on the results of my study, I called for extending our user-centered ethic of information and technology design to studies of IT adoption and use (Dayton, 2004b). To that end, I have begun to investigate whether and how our ideals about usability can be well served by SSCM, a technology that appears to be diffusing rapidly among technical communication work groups.

A focus on work groups is particularly important for understanding the rhetorical and sociotechnical dynamics driving the diffusion of SSCM, which is widely regarded as the most important new technology for technical communication since the advent of desktop publishing. This new method of creating technical publications radically changes traditional information-development processes. The IT most commonly linked with SSCM solutions is extensible markup language (XML), an open-source language that is being incorporated into a panoply of information and communication technologies, giving an aura of inevitability to SSCM. Although XML-based single sourcing has been hyped in the practitioner discourse of our field since the late 1990s, diffusion has been slowed by the widespread perceptions that SSCM solutions are too expensive, inflexible, unreliable, and require too much technical expertise to design and maintain. Those perceptions are rapidly changing. At the 2005 International Conference of the Society for Technical Communication (STC) in Seattle, I attended several standing-room-only sessions by work-group panels reporting successful implementations of SSCM. Each story was unique, and not all were entirely positive, but they all highlighted the message that SSCM was doable, affordable, and would deliver the return on investment that vendors promised, particularly if a company was doing any amount of localization of its information products. Dramatic cost savings were reported whenever localized translation into multiple languages was an integral part of the information development process.

AN ANALYTICAL FRAMEWORK TO STUDY IT ADOPTION AND USE

Each story of SSCM adoption is unique, and yet repeated patterns of setting, character, and plot can easily be discerned, along with a limited variety of themes. To organize multiple case studies of SSCM and to define the thematic patterns in them, we need a flexible, accessible, and adequate analytical framework.

Spinuzzi (2002) contended that constructing analytical frameworks to study and analyze collaborative work activity has multiple benefits, particularly when the frameworks focus on "interpretive issues that concern rhetoricians and technical communicators" (p. 98). With such frameworks, researchers pursuing similar goals can share a common conceptual vocabulary and develop standards to guide research design and analysis. The frameworks themselves can be improved as researchers critique and refine them. Like other influential researchers in workplace communication, Spinuzzi blended the conceptual and explanatory resources of North American genre theory and CHAT.

To study cases of IT adoption and use, I propose a framework blending CHAT with two others: (a) Rogers's (1962, 2003) ADT, and (b) the SCOT. In this section, I describe each of these theories and explain how I blended them into a hybrid analytical framework.

Organizing the Story: Rogers's ADT

If there were a specialized discipline called "Technology Diffusion Studies," its founding text would certainly have to be Rogers's (1962, 2003) *Diffusion of Innovations*. On the way to authoring the first edition, Rogers (1962) conducted a content analysis of 506 research reports from multiple disciplines, most of them carried out between the late 1930s and late 1950s. Rogers expanded and updated his theory over the decades, and the fifth edition (Rogers, 2003) claimed validation, with slight modifications to the original theory, from about 5,200 studies published in the literature of over a dozen disciplines.

I have summarized Rogers's (1962, 2003) ADT previously (Dayton, 2001, 2004b), and the reprise that follows glosses the innovation-decision process and the perceived characteristics of innovations—interrelated aspects of ADT that I extracted and revised for the analytical framework to study innovative IT adoption and use by work groups.

As depicted by ADT, a person or group deciding whether to adopt an innovation passes through five stages. In the first stage, knowledge acquisition, the key factors are the information available, the perception of a need for the innovation, and contact or lack of contact with change agents. During the persuasion stage, knowledge moves from general awareness to personal involvement and in-depth consideration, solidifying attitudes toward the innovation. The third stage culminates in a decision and often involves trying out the innovative tool or process. Implementation, the fourth stage, refers to incorporating or rejecting the innovation, and the final stage is confirmation that the decision was the right course of action.

Rogers's (2003) review of research on innovation adoption within organizations led him to describe a special version of the innovation-decision process. For organizations, the process often begins by focusing on a problem that is then matched with a potential innovation as a solution (Rogers, 2003, p. 421). The decision to adopt ushers in the implementation phase beginning with adaptation of the innovation to fit the organization, which usually results in some degree of alteration to the organization's structure. A further redefining of the innovation to clarify its place within the organization leads to the final stage of implementation, in which the innovation, having become completely normalized, loses its identity as an innovation.

In Rogers's (1962, 2003) theory, the innovation-decision process hinges on how the decision-making unit perceives the innovation with regard to these five characteristics:

1. How much of a payoff it seems likely to bring (relative advantage).
2. How easy or difficult it seems to fit with current processes (compatibility).
3. How easy or difficult it seems for users to understand and learn (complexity).
4. Whether, to what extent, and with what ease or difficulty it can be tried out (trialability).

5. How visible the results of adopting the innovation are likely to be (observability).

Rogers's (1962, 2003) theory has been validated mainly by analyzing cases of individuals adopting innovations for themselves. The framework has stood up well even in many studies of organizational IT adoption and use in which individuals have had the freedom to choose whether and to what degree to use the innovation. Within organizational contexts, however, additional social factors come into play, and they can vary dramatically across and even within organizations. Surprisingly, I have not been able to locate any published studies of collective IT adoption and use by work groups. When SSCM is the innovation being studied, however, the decision-making unit is often a technical documentation work group, as demonstrated by several of the cases I learned about from technical sessions of the annual STC conference in Seattle in 2005 (e.g., Bookless, Marx, & Davis, 2005).

Rogers's (1962, 2003) ADT offers a time-tested general framework for studying cases of IT adoption and use. When the focus is on the rhetorical dynamic of meaning making within a specific organizational context, this framework's innovation-decision model predicts that the adopting unit's meaning making will revolve around a relatively limited number of characteristics of the innovation, the most important of which will be relative advantage (usefulness), compatibility, and complexity (ease of use). The ability to try out the innovation and its salience to others may be important additional considerations.

Rogers's (1962, 2003) theory also predicts that case stories of IT adoption and diffusion by work groups will reveal a staged progression in the rhetorical dynamic: information awareness, seeking, interpreting, reframing, adapting, and appropriating. In the organizational version of the process, the process kicks into gear with the search for a solution to some problem.

Combining these two aspects of ADT, I have a simplified general model of what happens in a case of organizational IT adoption and diffusion, a model that fits a narrative template commonly used in academic as well as journalistic accounts. Though academic case studies typically thicken the narrative with conceptual rumination and reviews of prior theory and research, the structure of the typical case story is usually close to the following order of topics:

1. Description of the organization or work group prior to the innovation.
2. Description of the innovation, the perceived benefits versus drawbacks, and the decision making that led to its adoption.
3. Description of the resulting changes in the work group's social and technical characteristics.
4. Analysis of the rhetorical dynamic and sociotechnical factors shaping the adoption, implementation, and adjustment process, and the ultimate fate of the innovation within the organization or work group.
5. Recommendations regarding lessons learned and delimiters of applicability.

This case story template has proven adequate for studies of innovation adoption in many disciplines. However, to study the rhetorical dynamic and sociotechnical factors shaping the knowledge creation by work groups who are collectively adopting or have recently adopted a new IT tool, researchers need an analytical framework that allows them to probe more deeply and systematically into a group's innovation-decision process. For that purpose, CHAT, particularly as articulated by Engeström (1999a, 1999b), offers analytical tools that productively complement the aspects of Rogers's (1962, 2003) theory that I have just highlighted and simplified. Next, I explain the modifications of CHAT that I have integrated into a hybrid analytical framework for studying IT adoption and use.

Mapping the Context: CHAT

During the past decade, CHAT has informed a productive current of research and theory within technical and professional communication, influencing scholars such as Bazerman (1997, 2003), Haas (1996), Russell (1995, 1997, 1998), Spinuzzi (1996, 1999a, 1999b, 2002), and Winsor (1999, 2000). Kain and Wardle (2005) provided a comprehensive review of our field's appropriation of CHAT in a recent article in *TCQ*. In the interest of concision, the discussion that follows assumes some familiarity with the basic concepts of this theory.

In Leont'ev's (1978) original conceptualization of an activity system, a subject (one or more individuals) and an object are joined in a triangular configuration through the mediation of signs and/or tools, which the subject uses to transform the object into an outcome. The subject's collective motive is the key stabilizing feature of every activity system. The motive situates the subject intentionally with regard to the object, which is both the thing transformed by the activity and the activity's purpose, as in the object of the game. In Engeström's (1999a) expanded model, the subject's activity is depicted as embedded within a social context, its relation to a community of practice mediated by conventions, policies, and rules, whereas division of labor mediates the community's relationship to the object. Engeström's (1999a) model also broadens the concept of signs and/or tools to mediating artifacts.

CHAT is similar to Burke's (1969) pentad in its generative and heuristic function. The pentad's act, scene, agent, agency, and purpose provide an initial structure into an investigation into motives, which can then be shifted by analyzing one perspective as mediated by another. Fox (2002) quoted Burke's statement that "what we want is *not terms that avoid ambiguity*, but *terms that clearly reveal the strategic spots at which ambiguities necessarily arise*" (p. xviii; emphasis in original) and noted that "the pentad offers a method that reveals nuances, that exposes the messy and complicated nature of symbolic action" (p. 371).

CHAT provides us with an archetypal structure for analytical descriptions of work. Rather than draw attention to ambiguities, CHAT motivates analysts to seek out and explain the internal contradictions and sociocognitive dissonance embedded in collaborative work practices. Guided by CHAT, researchers are predisposed to look for disparities between what participants in an activity system describe as their goals and procedures and how the work actually gets done—as revealed by direct observation, documentary evidence, and the reports of other informants. Engeström liked to pepper his activity-system triangles with lightning-bolt images to mark the sources of breakdowns and incongruities within an activity system, which are regarded as "sources of development" (Kuutti, 1996, p. 34).

Engeström (1999a) viewed the normal, healthy evolution of activity systems in terms of "expansive cycles," a concept of collective learning derived from Vygotsky's zones of proximal development for individuals (p. 34). Engeström (1999b) presented a seven-stage expansive cycle of learning and knowledge creation by work groups. Wegner (2004) applied this chronological framework to describe the stages of a work group's construction of a management report. The seven stages parallel the stages of the innovation-adoption process for organizations described by Rogers (2003). Neither model, however, precisely describes the innovation-adoption processes by work groups that I have learned about with regard to SSCM. Rogers's stages overly generalize the first part of the process, whereas Engeström's (1999b) stages oversimplify the second part by making implementation a discrete point in the cycle of learning. The patterns of adoption I know about from several cases of SSCM adoption are more accurately described by considering implementation as a sequence of stages beginning when the work group commits to adopting a specific solution.

My framework to this point, then, combines Engeström's (1999a) expanded model of CHAT to describe a work group's activity system, setting the scene for the innovation-adoption story, whose plot is expected to play out in a sequence of stages that blend Engeström's (1999b) expansive learning cycle with Rogers's (1962, 2003) innovation-adoption process for organizations. This dual-theory framework can be further enhanced by zooming in for a real-time, close-up view of the rhetorical dynamic during the initiation part of the adoption process—before the innovation has become routinized or abandoned. Only participant–observer field studies can give us this in-depth look inside the black box of a work group's interpretive processes. To guide such studies, the SCOT offers an apt additional framework, which I cover next.

Opening the Black Box: SCOT

First proposed by Pinch and Bijker in a 1984 discussion paper that was reprinted as a shorter book chapter in 1987, SCOT enjoyed a brief heyday in science and technology studies in the 1980s and early 1990s. SCOT was pushed from the limelight by the more abstract and ontologically radical actor-network theory (ANT), which originated at about the same time. Both SCOT and ANT provide heuristic frameworks to guide empirical research aimed at uncovering and connecting up the rhetorical and sociotechnical interactions involved in the emergence, evolution, and stable definition (black boxing) of technological artifacts. Both theories trace their intellectual lineage to the strong program variant of the sociology of scientific knowledge (SSK) and the closely related empirical program of relativism (Sharif, 2005), analytical paradigms that shaped much of the work in the sociology of science during the 1970s and 1980s.

Bijker (1995) presented the most complete elucidation of SCOT in chronicling the three technologies mentioned in the title of his book *Of Bicycles, Bakelites, and Bulbs: Toward a Theory of Sociotechnical Change.* In brief, SCOT focuses on the unpredictable, open-ended *interpretive flexibility* of novel technologies, tracing their multidimensional and often widely divergent meanings to relevant social groups (RSGs) with an interest in the new artifact, technique, process, or system. Through discourse with one another and interaction with the technology, members of an RSG construct a *technological frame* that structures and eventually stabilizes the meaning of the technology for them. When technological frames clash, discursive wrangling among RSGs may interact with an ongoing process of technical tinkering and reinvention. The technology becomes subject to the vagaries of sociocognitive realignments triggered by both rhetorical and technological transformations.

SCOT depicts a new technology and the social groups deciding what it will mean as mutually adaptive; together, the technology and the social actors constructing its meaning form the sociotechnical ensemble, which is the unit of analysis in any given case (Bijker, 1995, p. 274). Typically, an analyst guided by SCOT deconstructs the mutually adaptive social and technical formation of the technology, tracing the interpretive conflicts among RSGs and the transformations in meaning that emerge from a process blending adjustments that are both technical and interactional and rhetorical and perceptual. The social construction of the technology reaches closure when the interpretations of the RSGs converge, and the process fades into collective amnesia when the technological innovation reaches stabilization.

SCOT's originators grounded their theory in cases of technology diffusion at the societal level; however, the theory's constructs can be adapted to examine cases of technology adoption within even small work groups by focusing on technological frames directly. After all, the purpose of focusing on RSGs when tracing diffusion at the societal level is to document the interpretive flexibility in the evolving technological frames of groups competing to dominate the stabilized meaning of the technology.

Summary and Review: The Complete Hybrid Framework

Denzin (1978) applied the concept of triangulation to four elements of a research study: data, methodology, investigator, and theory. The least explored of these is theoretical triangulation, in which an analyst applies multiple theoretical frameworks to construct an interpretive account that produces more insights and questions for further study than a single theoretical perspective would generate. Theories are useful to the extent that they help us organize and make sense of patterns we find in empirical observations, but they are selective in what they let us see. Combining theoretical perspectives in complementary and/or contrastive analyses can open our eyes to additional connections and patterns, bringing other dimensions into view. Researchers of workplace communication practices, for example, have demonstrated the added explanatory power achieved by combining CHAT and genre theory.

Researchers investigating cases of IT adoption and use by technical communication work groups will benefit by combining the three theoretical frameworks I have discussed. CHAT can help an analyst map the sociotechnical context and rhetorical dynamic of an activity system before and after the introduction of an innovation. The CHAT framework is particularly helpful in identifying triggers of the expansive learning cycle that set the stage for innovation and prime the community to seek or impose a change in the mediating artifacts, the object, the outcome, or even in the system's underlying motive. CHAT can also help uncover contradictions and breakdowns in the postinnovation activity system, which may be unacknowledged or discounted by some in the work group.

Engeström's (1999b) model of the expansive learning cycle blended with Rogers's (2003) description of the innovation-adoption process for organizations provides a paradigmatic plot that can add clarifying structure to a particular work group's experience in searching for, learning, interpreting, testing, and adapting a new IT tool or system. Rogers's (1962, 2003) perceived characteristics of innovations add a framework for collecting, organizing, and analyzing opinions and observations to deepen insights into the work group's assessment of the innovation and its impacts.

Finally, SCOT can deepen the understanding of a work group's innovation-adoption process by structuring a close-up view of the interpretive drama forming and reforming and finally stabilizing the group's perceptions of the innovation. This requires adequate, first-hand observations of the group's learning and communication process generated by collaborating and competing technological frames, including the communications of key decision makers interacting with the group and with vendors' representatives involved in marketing, training, and support.

Table 1 provides a complete summary view of the hybrid framework I am proposing to guide the study of innovative IT adoption and use by work groups. In the next section, I apply this framework in telling the story of a small work group in a mid-size high-tech company that implemented a SSCM system in 2004.

TABLE 1
Summary View of the Complete Hybrid Analytical Framework

Cultural-Historical Activity Theory (CHAT)	
Conceptual Lenses	*Focal Points of Questioning, Analysis, and Interpretation*
Activity System	
I. Sociotechnical Context and Ensemble	Culture and history or the organization and the work group
Subject and community	Hierarchy, division of expertise, and work roles and
mediated by rules and	responsibilities within the work group
norms that govern	Status of work group and degree of autonomy
interaction within group	Chain of command within the organization
and with other groups	Interrelationships with other work groups
Object(ive) and	Quality of leadership and innovativeness, creativity
community mediated by	Other characteristics contributing to unique sociotechnical context
division of labor	Character and closeness of contact with customers, awareness of
Motive and desired outcome	their needs, priorities, concerns
mediated by object(ive).	
II. Subject and Object(ive)	Group view of activity before innovation versus group view after
Mediated by Tools,	Views of individual group members before and after innovation
Artifacts, Procedures	Contradictions, dissonance, and breakdowns—acknowledged
	and unacknowledged—before and after innovation

Expansive Learning Cycle and Rogers's Organizational Adoption Process	
Initiation	Questioning > analyzing > modeling > testing
Implementation	Committing > adapting > re-defining > routinizing or
	abandoning

Rogers' Adoption and Diffusion Theory (ADT)	
Perceived Characteristics of Innovation	Relative advantage / usefulness
	Compatibility / fit
Group members' opinions	Complexity / ease of learning and use
regarding aspects of the	Trialability
innovation	Observability (by others)/visibility of benefits
	Adaptability (added to Rogers's five characteristics)

Social Construction of Technology (SCOT)	
Interpretive Flexibility	Rhetorical analysis of the innovation's organizing vision and
	variations thereof communicated by marketing material and
	vendors of competing products
	Rapid evolution of decision makers' interpretation of the
	innovation as they process and reconfigure information taken in
Technological Frames	Predisposing orientations toward innovation by key group
	members and decision makers, other group members
Closure and Stabilization	Rhetorical dynamic in negotiation, construction, re-construction
	of the meaning of the innovation by the group, culminating in
	"a new normal" for mediation of work activities and resulting
	changes in the sociotechnical context

IMPLEMENTATION OF SSCM AT AUTOMATED LOGIC

The site of this case study was the headquarters of Automated Logic Corporation (ALC), which makes integrated hardware, software, and firmware to control building systems: HVAC (heating, ventilation, and air conditioning), lighting, fire-sprinkler, security, and other automated systems. The tagline on ALC's website is "Innovative building controls solutions made powerfully simple" (www. automatedlogic.com). Its newest product offers full control of building systems over the Web.

I learned about ALC's implementation of SSCM through a technical writer, Bev Arends, who e-mailed me in response to an announcement I posted to the listserv of technical communication alumni from Southern Polytechnic State University in Marietta, Georgia, where I was a faculty member. The announcement stated my desire to interview technical communicators to learn more about their experiences using SSCM systems. Bev passed my request to Chariti Young, the leader of the knowledge management (KM) group at ALC. At Chariti's invitation, I interviewed her by phone, recording the conversation and taking notes while we talked. I visited ALC headquarters a few days later, spending nearly three hours with Chariti and Todd Cantrell, coleader of the KM group in charge of the technical support website. They demonstrated the website's graphically rich "information access user interface," the development of which—along with the database information architecture behind it—was the most visible result of the team's 12-month project to implement an SSCM system. During this meeting at ALC, Chariti and Todd also opened AuthorIT to give me a look at some of the functions and procedures they use to develop and maintain their SSCM system. They also showed me Microsoft Excel spreadsheets as they explained how they performed a content audit, having used Excel to construct a content model early in the project.

As a final step in data-gathering for this study, I interviewed technical writers Bev Arends and Cindy Pickrell, recording the conversations so that I could go over what they said and develop a comprehensive summary. I asked for and received additional information from Chariti by e-mail in the form of a timeline outlining the milestones in their SSCM adoption project. Finally, Chariti also read my journalistic draft account describing the case and returned comments to clarify and/or correct some points. She also provided a few suggested changes in wording, all of which I accepted. Drawing text directly from that revised account, I next present a reorganized version that adds observations and interpretations suggested and shaped by concepts from the hybrid analytical framework.

Sociotechnical Context and Ensemble

ALC is headquartered in Kennesaw, Georgia, a town of about 30,000 on the northwestern fringe of metro Atlanta. In business since 1977, the company experienced robust growth over the past decade, increasing from 60 employees in 1996 to about 250 in 2004 when it was purchased by Carrier Corporation. ALC's first-line customers are its network of over 150 dealers, independently owned companies that design, install, and maintain automated building-control systems.

Carrier's acquisition of ALC presents a contrast in organizational profiles—and in the cultures that go along with them. Carrier, a subsidiary of United Technologies Corporation, employs more than 45,000 people and has 80 manufacturing plants spread over six continents. Whereas Carrier is a paragon of corporate hierarchy and organizational complexity, ALC embodies the flattened and simplified organizational structure of small and mid-size high-tech companies. ALC's headquarters is a one-story building with more than 50,000 square feet of open space divided into cubicles. Plain but attractively lettered rectangular signs hang from the ceiling, marking the general location of the company's work groups: *Documentation*, *Marketing*, *Technical Support*, and so forth. Inspirational messages are painted in large cursive writing high on the walls surrounding the cubicle-maze. There are no corner offices or offices of any type, and it is not possible to discern a hierarchical arrangement to the cubicles, which all look pretty much alike. There are, however, ample and well-equipped conference rooms.

The innovator in this story was a work group composed of two knowledge managers, three technical communicators, and a Web developer.[1] The group's leader, and my principal informant, was Chariti Young, who had been working for ALC for about five years when I made contact with her in 2005. Chariti led the KM effort at ALC with Todd Cantrell, a long-time ALC employee who came to KM by way of technical support. Both Chariti and Todd are engineers, she with bachelor of science and master of science degrees in mechanical engineering from Penn State and he with a bachelor of science degree in electrical engineering technology (communications option) from Southern Polytechnic State University.

The three technical communicators in ALC's SSCM development group were Pat Arterburn, Cindy Pickrell, and Bev Arends. Pat transferred from ALC's software-testing department and began technical writing at about the same time the group began searching for an SSCM solution. Cindy had over 20 years of experience in technical writing at the time of this story, and her career spans the typical trajectory of the key technologies in our field, from typewriters to word processors to FrameMaker and Robohelp, and, finally, to AuthorIT. Bev Arends came to technical writing later than Cindy, by way of a master's degree in technical and professional communication from Southern Polytechnic. It was Bev who first made contact with me in response to my call for informants about SSCM systems. Bev, it turned out, played a key knowledge-transfer role early in the group's search for information about SSCM.

---

[1]Chariti and the other employees of ALC mentioned in this article consented to participate in the study and approved the use of their real names.

Web programming was critical to the success of that system; indeed, improving the usability of Web-delivered content was the primary focus of the group's search for an information-development system. As a result, Brian Green, ALC's Lotus Notes guru and Web programmer, also played a pivotal role in evaluating content management alternatives and adapting AuthorIT to meet the special publishing needs of ALC.

TRACING THE INNOVATION-ADOPTION PROCESS

Chariti first discovered the applicability of single sourcing at ALC when she was put in charge of the training group and wanted to use content that had been created by the technical writers in the documentation group. She had hired on at the company as an engineer in mid-2000 and spent nearly two years in the engineering group. In previous engineering positions, she had developed something of a secondary specialty in training, and she welcomed the opportunity in 2002 to develop Web-based training (WBT) at ALC. She was later put in charge of both training and documentation, and this was when she discovered opportunities for single sourcing content at lower than the document level. For example, Chariti discovered what appeared to be the same piece of technical information worded five different ways in as many documents. It took a meeting of engineers and technical writers to figure out which of the chunks varied only by style and which had distinct meanings.

The documentation and training work groups used FrameMaker and Web-Works Publisher to create both portable document format (PDF) manuals and hypertext markup language (HTML) pages published as electronic product help and on the ALC support website. Under Chariti's leadership, the content developers in these groups came up with some makeshift single-sourcing procedures, "workarounds that didn't work really well." Nevertheless, the awareness of single-sourcing opportunities prepared the group to begin thinking in terms of content objects at lower than the document level.

Using a home-grown Lotus Notes application, technical support engineers at ALC captured knowledge continuously generated by their problem-solving and troubleshooting roles. Their Lotus Notes application was an efficient document management system, but the content it generated was not granular enough to be mined efficiently for reuse and repurposing by those in the documentation and training work groups.

Questioning and Analyzing

In January 2004, around the time of the company's acquisition by Carrier, ALC underwent downsizing and restructuring. The training and documentation groups went from a total of 12 employees to 7 and merged with the technical support group under the supervision of Michael Facente, head of technical support. "The challenge to do more with less is always a challenge that will squeeze you into places you wouldn't necessarily have gone before," Chariti told me.

The organizational restructuring happened at a time when the technical support group had begun to focus on revamping the website for ALC's dealers—the 150-plus companies around the world that use ALC's products to design, install, and maintain automated building-control systems. The company knew its support website had to be overhauled; persistent complaints from dealers had made that necessity clear. As Michael, Chariti, Todd, and the remaining technical communicators took stock of the situation, it also became clear that the solution would involve object-oriented information development, which was how Chariti and the other engineers on the team thought of SSCM. Building a document from a database of information chunks is similar to object-oriented programming, in which developers draw from libraries containing modular units of code.

As part of the restructuring, Chariti was given the lead KM role, which involved "looking at the big picture of documentation, looking at where all of our content comes from and where it goes," she said. She found support and direction in books recommended by Bev Arends, who also advocated a user-centered design approach to revamping the support website. The two projects were parallel and complementary: to find and implement an appropriate content management system and to develop a user-friendly website serving up all the company's technical information dynamically in search-generated hypertext collections as well as in libraries of PDF manuals.

Modeling, Testing, Committing

Two books were critical resources for the team: Rockley's (2002) *Managing Enterprise Content: A Unified Content Strategy* and Goto and Cotler's (2001) *Web ReDesign: Workflow That Works.* Following the content-audit process explained in Rockley's book, the team began by analyzing the content of the knowledge-base website. Soon after that task was underway, Chariti traveled to Toronto to attend a Rockley workshop on implementing a unified content strategy. She was able to use her just-started audit in the workshop, receiving suggestions and tips from Rockley and workshop participants. She returned to ALC headquarters in Georgia energized by greater confidence in her knowledge of the content-modeling process and by new information and ideas about SSCM solutions. She put the three technical writers in charge of creating a content model, beginning with an inspection of the entire documentation and training library to identify which information chunks could be eliminated and which could be reused for multiple products and outputs. They also looked for standard elements of topics that could be written once and embedded in multiple topics. The same process was applied to the content on the technical support website.

Early in this process, the team began to receive visits from vendors of content management systems, document management systems, and even e-learning content management systems. They soon realized that most of these systems were more expensive and more complex than ALC needed, and they did not offer well-integrated options for the three publication outputs from the same content database that ALC wanted: PDF manuals, context-sensitive help created in extensible HTML (XHTML), and support website pages also created in XHTML in response to user search queries. At the Rockley workshop, Chariti learned of another SSCM solution she had not been aware of previously: AuthorIT, a software application designed particularly for single-sourcing documentation. This product was less expensive than other solutions ALC had considered; it appeared to be easier to install and learn than most; and it provided the publishing options ALC needed—with some work-arounds. Trying out AuthorIT in its 30-day free trial convinced Chariti and Todd that they had found the best solution.

Adapting and Redefining

During the initiation phase of the innovation-decision process, ALC management's desire to overhaul the company's technical support website became the dominant driver in the search for an SSCM solution. ALC's dealers had been complaining for years about the difficulty of finding information on the support website. ALC had previously executed two radical redesigns of the site that had failed to meet the dealers' needs. When ALC management gave the technical support team a do-or-die charge to fix the website, Chariti's KM team decided to implement the user-centered design process presented in Goto and Cotler's (2001) book mentioned earlier. The technical support engineers identified the most outspoken complainers about the website, and technical writers Bev and Cindy interviewed 40 of them by phone.

The interviews gave the team a starting place for the redesign. The information from users verified some of the team's assumptions and negated others about what needed to change on the support website. From the interview data, the team constructed personas for two primary user groups. They also conducted a group card-sort exercise with the technical support engineers, having them sort and label the high-level categories of information in the knowledge base. From the results of that card-sort, the team collaboratively designed the information architecture and built a prototype site. They iterated through several rounds of usability tests using HTML mock-ups of top-level search and results pages.

In redesigning the site, the team's highest priority was communicating to users the options for selecting appropriate filters in conducting a search of the knowledge base. The user research made the team confident that their prototype design would meet users' needs, but initial usability tests revealed that actual users from the dealerships could not easily make sense of the new options for searching the support website. This led the team to make significant changes in the information architecture and, just as important, in the information design—the arrangement and wording of the user-interface elements. The most significant change in the latter was the decision to use screen-shot graphics to cue search-filter categories in the website's user interface. Technical writer Bev Arends presented this idea and spearheaded the development of the customized screen-shot components, which were then verified by further usability testing.

The key technical requirement that emerged from the usability tests was that information in the knowledge base would need to be tagged by category and information type to enable greater search-filtering efficiency. Accomplishing this additional metatagging was not a straightforward process in AuthorIT, and it took the team a little while to discover a work-around.

AuthorIT was well suited to the needs of the documentation and training team, and the team members particularly liked the keyword-tagging function, which enabled them to quickly add multiple keyword metatags to content elements. When they realized they also needed to add metatags for products, functions, and information types (procedural, conceptual, trouble-shooting, etc.), they did not immediately find a function in AuthorIT that suited their needs. The team's queries to AuthorIT technical support and to relevant online support groups did not produce a satisfactory answer. It was only from looking under the hood of AuthorIT that the team members discovered they could use the variable metatags to accomplish what they needed to do.

The team was able to stretch AuthorIT's standard functionality in several other ways that were critical to meeting all of ALC's specific publishing needs. AuthorIT offers a number of stock output types, including XHTML, but for each output type, AuthorIT allows only one template. The team members needed to produce two different flavors of XHTML, one for the website and for one context-sensitive help, so they came up with a batch file that copied the XHTML template needed at any given publishing moment, replacing AuthorIT's stock template.

The team also had to tweak AuthorIT's stock process to produce PDF manuals, which is done through Microsoft Word. Coding topic identifiers into the online help files required another work-around, which Chariti implemented in JavaScript. In a fifth noteworthy adaptation of AuthorIT, this one suggested by the product's sales representatives, the KM team customized AuthorIT's Localization Manager to rebrand information products for partner firms marketing hardware and software products with their own company names, product names, and logos.

Routinizing, With Minor Reservations

ALC was quite satisfied in the summer of 2005 with its customized SSCM system based on AuthorIT, a structure query language (SQL) database, and the Verity search engine that parses the content metatags used for filtering searches on the support website. The website had garnered accolades from users, and the entire KM team was proud of what they had accomplished and comfortably settled into a new, more productive routine—doing more with less.

Chariti mentioned two residual concerns. One was that the KM team's use of variable metatags was "really a stretch from what AuthorIT envisioned," Chariti said. Would future versions of AuthorIT incorporate changes that would make the current work-arounds unworkable? "Because it is XML," Chariti said, "we feel pretty confident that even if AuthorIT is bought out by somebody bigger ... we'll still be able to use it for what we're doing."

The manual awkwardness of the make-do procedures they had devised to achieve critical functionality was the other concern. When the AuthorIT representative learned how ALC was using the variable metatags, he arranged for some AuthorIT developers to visit ALC for a detailed explanation of the work flow and work-arounds Chariti's team had come up with. The AuthorIT visitors were much impressed and were slated to make a return visit. Chariti hoped that AuthorIT Software Corporation would make it a goal to work her team's innovations into their product as integrated and more automated functions.

SUMMARY VIEW, CHAT PERSPECTIVE

I find a ready fit between CHAT's key terms and specific aspects of ALC's conversion of its multiple-tools, multiple-outputs process to an SSCM system using AuthorIT. For starters, Christi and others in the KM group at ALC became aware of a contradiction in their former activity system when Chariti wanted to mine documentation products to develop WBT modules. In CHAT terms, the creation of a new activity system that interacted with an existing one led to the awareness that ALC's procedures for creating documentation produced redundant content that spread, untracked, through multiple documents. As a result, the repeated content elements could not easily be updated and edited for accuracy and consistency, and they could not easily be found for reuse and repurposing. Frustration about this inefficiency motivated Chariti and her colleagues to begin the inquiry into a single-sourcing alternative for information development. They developed more efficient procedures using FrameMaker and WebWorks Publisher, but the motivation and justification for moving to the object-oriented paradigm of SSCM was lacking—until the company imposed a new, top-priority motive.

In CHAT, the motive underlying any activity system situates the subject intentionally with regard to the object, which is both the thing transformed by the activity and the activity's purpose, as in the object of the game. When ALC was acquired by Carrier, the larger activity network imposed a new, top-priority motive: Do more with less. This sparked a reengineering of the activity systems producing documentation, training, and the Web-delivered knowledge base: Three related work groups merged into the KM team, and the division of labor expanded to include content-analysis and content-database development and maintenance. The new motive imposed by downsizing and restructuring provided the impetus for Chariti and her newly reconstituted team to take steps rapidly to reengineer their activity system around a new object—a single-source database for managing all the content generated by those working in documentation, training, and technical support.

In the new activity system that emerged, the chief mediating artifact became AuthorIT, which enabled the KM team to create and maintain the componentized content database and to publish, in multiple formats and media, the desired outcome: technical information for ALC's dealerships that was more complete, accurate, useful, and easily findable.

ADT: INTERPRETING THE COST-BENEFIT PERCEPTIONS

In adopting AuthorIT, adapting it to their needs, and making its use the new routine mediation of their work, the KM work group at ALC constructed perceptions of AuthorIT that reflected a highly positive rating on each of the five characteristics identified by Rogers's (1962, 2003) ADT as decisive criteria in the adoption-decision process.

Relative Advantage

After the documentation, training, and technical support groups were merged, the leaders of the new KM group became convinced that developing an SSCM system was the only way they could meet their goals for improving the usability of the support website and generally working more efficiently. Thus, their consideration of relative advantage during the early stages of the innovation-assessment process focused on comparing products and solutions from competing vendors. In evaluating acquisition, training, and transition costs, they deemed AuthorIT to offer a significant advantage relative to the other products they evaluated. AuthorIT was much less expensive than the most attractive Web-centered solutions they examined, but it also won out when they evaluated it in terms of the other four perceived characteristics of innovations.

Compatibility

The KM team naturally hoped to find a system that promised to lower the costs and alleviate the difficulties of changing from processes built around FrameMaker for manuals, Webworks for online help, and XHMTL for the support website. They were skeptical they could find a tool that would be a good fit for developing all three types of information products. Many of the content management systems they looked at were Web-centered and would require giving up their homegrown Web development tools and procedures. Few vendors had feasible solutions for dealing with ALC's need for print manuals and context-sensitive help. AuthorIT, on the other hand, offered publishing options that were a close, though not perfect, fit with the three types of information products that ALC wanted to continue publishing in forms close to what they already had. A consideration that weighed heavily in their decision to go with AuthorIT was that it offered a relatively painless conversion process for FrameMaker files.

Complexity

AuthorIT also impressed the KM team because it seemed simpler than the other tools on their short list. A key component of AuthorIT's positioning in the marketplace is that it hides complexity from users. Posting to the STC's Single-Sourcing listserv recently, the head of AuthorIT Software Corporation wrote, "One of the great things about our approach is that AuthorIT exposes the concepts and methodology of single-source content management without exposing all [the] technology that is used to implement it, which can confuse and over complicate things" (Trotter, 2005). ALC's technical writers, Bev and Cindy, reported that they had found AuthorIT relatively easy to learn and very easy to use, while expressing awareness that the tool was much more complex than they had to be concerned about, given their roles. They were glad to have Chariti and Todd master the more complex levels of managing the AuthorIT system.

Trialability

The KM team downloaded the free trial of AuthorIT and tried it out before deciding to purchase. The group had specific technical questions about whether AuthorIT would do everything they needed to accomplish with it and how easy or difficult it would be to learn. Their trial use of AuthorIT reinforced their belief that they could get it to do everything they needed it to do.

Observability

The degree to which an innovation is anticipated to attract attention and approval is the fifth perceived characteristic influencing innovation adopters. The KM team's primary goal, set by upper management, was the redesign of the support website. The team's successful adoption of SSCM and their implementation of a user-centered design process enabled them to achieve their goal, whose effects have been highly visible to management, dealers, and the technical support team of engineers.

Adaptability as a Sixth Criterion

Usability testing led the group to realize that search filtering by product, function, and information category would be essential to make the improvements necessary to the support website. At first, the group was not sure their recently purchased solution could do what they now realized it needed to do. The group had to stretch AuthorIT, finding a way to implement the multiple metatagging they needed to support the search filtering to be implemented on the website. This and the other adaptations of AuthorIT mentioned earlier illustrate a concept added to the most recent version of ADT—reinvention (Rogers, 2003). In the 1970s, research in innovation diffusion began to look at cases farther along in the implementation stage and found that reinventing an innovation to fit particular needs was common. It probably strikes most of us as common sense, so accustomed have we become to finding just-in-time, ad hoc ways to get work done with new computer and communications tools.

WHAT SCOT COULD ADD TO THE PICTURE

The process of reconfiguring its activity system to incorporate SSCM added an important chapter to the ALC work group's cultural history, an account of a period that Chariti compared to "being in the middle of a tornado,"—in other words, a survival story. She recalled the collaborative learning and problem solving the team went through as inflicting "a lot of brain damage." Initially, there was the problem of sorting out and analyzing what the team needed versus what vendors of content management systems were selling.

In the process of evaluating SSCM solutions presented to them by over a dozen vendors, ALC's decision-making group included at least three potentially distinct technological frames. The primary technological frame was the one developed by the technical support engineers and the Web developers responsible for the support website. Making that website more usable—with the goal, eventually, of getting customers to use the website before calling technical support—was the overriding goal of the effort to implement SSCM. Probably due at least in part to the emphasis on the support website, the ALC group focused a lot of effort evaluating Web-centered content management systems.

A second technological frame was represented by the technical writer of a spin-off company housed in a separate nearby building. That technical writer would need to work with the content database created by ALC's KM team to create manuals and online help rebranded with product names, descriptions, and logos of the associated company. His primary requirement for the SSCM, then, was that it should enable him to work with the same content components produced by his colleagues in ALC and use conditional variables to output and maintain a customized documentation library. He was included as a key player in the evaluation of the solutions proposed by vendors, and Chariti and Bev both mentioned the importance of making sure the solution would be compatible with his needs.

The ALC technical writers represent a third technological frame in the company's evaluation of SSCM solutions. Bev and Cindy said they were not active participants in the evaluation of alternatives, although each of them attended at least one of the vendor presentations. Both of them said they were not concerned about which product the engineering-led team would choose, trusting that whatever authoring–editing tool was selected would be well within their ability to learn. They were pleased at the choice of AuthorIT because it was relatively easy to learn. It allowed a division of labor in which they did not have to concern themselves with the deeper technical knowledge that Chariti and Todd had to develop to make the whole system work.

In the ALC case, interpretive flexibility among technological frames came into play most intensively in interactions with the technological frames of individual product vendors. Unfortunately for a deeper examination of this case, SCOT predicts that the complex rhetorical process involving requirements, claims, questions, demonstrations, and internal team discussions will be nearly impossible to analyze when there is no one to observe and record the flow of communication as it happens. Once closure and then stabilization have set in after a selected product has been implemented, the interpretive matrix of definitions and decisions leading to the choice are no longer readily accessible to examination. In the case of an organizational work group, the retrospective accounts of key actors of the sociotechnical ensemble are likely to be incomplete and influenced by the intervening experiences with the product that was implemented. I expect it would be possible for Chariti, Todd, and others at ALC to reconstruct a general account of their thinking about the few products they were presented with that seemed to be viable solutions. However, what would be really useful to have is a detailed replotting of the sociocognitive drama that unfolded slowly over several months as the team's multiple technological frames converged while being informed by and compared with the technological frames of vendors. By its very nature, that discursive activity can be adequately and reliably recorded only by observers close to the action while it is going on or soon afterward while the memories of key participants are still fresh.

SUMMARY AND CONCLUSION

In this article, I have rearticulated the perspective argued by Doheny-Farina (1992) that "technological innovations are part of complex social, organizational, institutional interactions, interpretations, and negotiations" (p. 7). This quintessentially rhetorical process "engenders a reciprocal shaping as it develops; the innovators, the innovation, and the users of the innovation are all changed through the process" (p. 6). I have advocated that we extend our long-standing interest in the rhetorical dynamic of technology transfer and diffusion to studies of innovative IT adoption and use by work groups, a focus that has generally been neglected both in our research and in research from the interdisciplinary domain of information science, technology, and management.

SSCM is still in the early stages of technology transfer and diffusion. Academics in technical communication should ally themselves with practitioners in interpreting this technology in ways consistent with our field's user-centered focus and our desire to persuade our colleagues in management and IT of the value we can add to strategic business goals (Clark, 2002; Hart-Davidson, 2001). I have proposed that we extend our user-centered ethic from information and technology design to innovative IT adoption and use (Dayton, 2004b). SSCM is a logical starting place. We need to know much more about the varieties of SSCM solutions, the advantages and trade-offs that work groups are negotiating as they enter expansive learning cycles in search of implementations that will fit best into their particular activity systems. The hybrid framework I have explicated can guide invention, information gathering, and analysis in studies of SSCM adoption and use. Its layered lenses can focus analysts' observations of the sociotechnical context, ensemble, and cultural history; structure the story of the innovation search, adoption, adaptation, and routinization or abandonment; and guide the close-up study of groups' intensive, kaleidoscopic interpretations and negotiations during the early stages of the adoption process.

By focusing our research and theory more actively on this important convergence of rhetoric, innovation, and technology, we can become more persuasive participants in the knowledge creation of practitioners who are compelled to come to terms with this new way of creating technical information products. And beyond SSCM, the sites for research into IT adoption and use are manifold; they offer opportunities to bring our critical, interpretive approaches and our user-centered ethic into contact with other disciplines focused on learning how to make the innovation-adoption process more successful—in social as well as in technical and economic terms.

REFERENCES

Bazerman, C. (1997). Discursively structured activities. *Mind, Culture and Activity, 4,* 296–308.

Bazerman, C. (2003). Speech acts, genres, and activity systems: How texts organize activity and people. In C. Bazerman & P. Prior (Eds.), *What writing does and how it does it: An introduction to analyzing texts and textual practices* (pp. 309–340). Mahwah, NJ: Lawrence Erlbaum Associates, Inc.

Bazerman, C., & Russell, D. (Eds.). (2002). *Writing selves/writing societies: Research from activity perspectives.* Retrieved October 3, 2005, from http://wac.colostate.edu/books/selves_societies/

Beason, G. (1996). Redefining written products with WWW documentation: A study of the technical writing process at a computer company. *Technical Communication, 3,* 339–348.

Bijker, W. E. (1995). *Of bicycles, Bakelites, and bulbs: Toward a theory of sociotechnical change.* Cambridge, MA: MIT Press.

Blomberg, J. L. (1986). The variable impact of computer technologies on the organization of work activities. In *Proceedings of the 1986 ACM Conference on Computer-Supported Cooperative Work, Austin, Texas, December 3–5, 1986* (pp. 35–42). New York: ACM Press.

Bookless, J., Marx, A., & Davis, S. (2005). Demonstration of an XML-based content management system. In *Proceedings of the 52nd International Conference of the Society for Technical Communication* (pp. 187–190). Arlington, VA: Society for Technical Communication.

Burke, K. (1969). *A grammar of motives.* Berkeley: University of California Press.

Clark, D. (2002). Rhetoric of present single-sourcing methodologies. In *Proceedings of the 20th annual International Conference on Computer Documentation* (pp. 20–25). New York: ACM Press.

Dayton, D. (2001). *Electronic editing in technical communication: Practices, attitudes, and impacts.* Doctoral dissertation, Texas Tech University, Lubbock. Retrieved October 3, 2005, from http://pages.towson.edu/ddayton/

Dayton, D. (2003). Electronic editing in technical communication: A model of user-centered technology adoption. *Technical Communication, 50,* 192–205.

Dayton, D. (2004a). Electronic editing in technical communication: The compelling logics of local contexts. *Technical Communication, 51,* 1–16.

Dayton, D. (2004b). Electronic editing in technical communication: A model of user-centered technology adoption. *Technical Communication, 51,* 207–223.

Denzin, N. K. (1978). Strategies of multiple triangulation. In N. K. Denzin (Ed.), *The research act: A theoretical introduction to sociological methods* (2nd ed., pp. 297–313). New York: McGraw-Hill.

Doheny-Farina, S. (1992). *Rhetoric, innovation, technology: Case studies of technical communication in technology transfers.* Cambridge, MA: MIT Press.

Downing, J. R. (2004). "It's easier to ask someone I know." *Journal of Business Communication, 41,* 166–191.

Duin, A. H., & Hansen, C. J. (Eds.). (1996). *Nonacademic writing: Social theory and technology.* Mahwah, NJ: Lawrence Erlbaum Associates, Inc.

Engeström, Y. (1999a). Activity theory and individual and social transformation. In Y. Engeström, R. Miettinen, & R. Punamäki (Eds.), *Perspectives on activity theory* (pp. 19–38). Cambridge, England: Cambridge University Press.

Engeström, Y. (1999b). Innovative learning in work teams: Analyzing cycles of knowledge creation in practice. In Y. Engeström, R. Miettinen, & R. Punamäki (Eds.), *Perspectives on activity theory* (pp. 377–406). Cambridge, England: Cambridge University Press.

Fichman, R. G. (2000). The diffusion and assimilation of information technology innovations. In R. Zmud (Ed.), *Framing the domains of IT management: Projecting the future through the past.* Cincinnati, OH: Pinnaflex Educational. (Also available from http://www2.bc.edu/~fichman/Fichman_2000_IT_Diff_Chp.pdf/)

Foucault, M. (1972). *The archaeology of knowledge* (A. Sheridan, Trans.). New York: Pantheon.

Fox, C. (2002). Beyond the "tyranny of the real": Revisiting Burke's pentad as research method for professional communication. *Technical Communication Quarterly, 11,* 365–388.

Freedman, A., & Medway, P. (Eds.). (1994). *Genre and the new rhetoric.* London: Taylor & Francis.

Goto, K., & Cotler, E. (2001). *Web redesign: Workflow that works.* Indianapolis, IN: New Riders.

Haas, C. (1996). *Writing technology: Studies on the materiality of literacy.* Mahwah, NJ: Lawrence Erlbaum Associates, Inc.

Hansen, C. J. (1996). Contextualizing technology and communication in a corporate setting. In A. H. Duin & C. J. Hansen (Eds.), *Nonacademic writing: Social theory and technology* (pp. 305–324). Mahwah, NJ: Lawrence Erlbaum Associates, Inc.

Hart–Davidson, W. (2001). On writing, technical communication, and information technology: The core competencies of technical communication. *Technical Communication, 48,* 145–155.

Kahn, R. (2000). The effect of technological innovation on organizational structure: Two case studies of the effects of the introduction of a new technology on informal organizational structures. *Journal of Business and Technical Communication, 14,* 328–347.

Kain, D., & Wardle, E. (2005). Building context: Using activity theory to teach about genre in multi-major professional communication courses. *Technical Communication Quarterly, 14,* 113–139.

Kramer, R. (2003). IBM's SGML toolset and the writer as technologist, problem solver, and editor. *Technical Communication, 50,* 328–334.

Kuutti, K. (1996). Activity theory as a potential framework for human–computer interaction research. In B. A. Nardi (Ed.), *Context and consciousness: Activity theory and human–computer interaction* (pp. 17–44). Cambridge, MA: MIT Press.

Lapointe, L., & Rivard, S. (2005). A multilevel model of resistance to information technology implementation. *Management Information Systems Quarterly, 29,* 461–491.

Leont'ev, A. N. (1978). *Activity, consciousness, and personality* (M. J. Hall, Trans.). Englewood Cliffs, NJ: Prentice Hall.

Meyer, A. D., & Goes, J. B. (1988). Organizational assimilation of innovations: A multilevel contextual analysis. *Academy of Management Journal, 31,* 897–923.

Orlikowski, W., & Yates, J. (1994). Genre repertoire: Examining the structuring of communicative practices in organizations. *Administrative Science Quarterly, 39,* 541–574.

Orlikowski, W. J. (1992). The duality of technology: Rethinking the concept of technology in organizations. *Organization Science, 3,* 398–427.

Orlikowski, W. J., & Baroudi, J. J. (1991). Studying information technology in organizations: Research approaches and assumptions. *Information Systems Research, 2*(1), 1–28.

Orlikowski, W. J., & Gash, D. C. (1994.) Technological frames: Making sense of information technology in organizations. *ACM Transactions on Information Systems, 12,* 174–207.

Pinch, T. J., & Bijker, W. E. (1984). The social construction of facts and artifacts: Or how the sociology of science and the sociology of technology might benefit each other. *Social Studies of Science, 14,* 399–441.

Pinch, T. J., & Bijker, W. E. (1987). The social construction of facts and artifacts: Or how the sociology of science and the sociology of technology might benefit each other. In W. E. Bijker, T. P. Hughes, & T. J. Pinch (Eds.), *The social construction of technological systems: New directions in the sociology and history of technology* (pp. 17–50). Cambridge, MA: MIT Press.

Porter, J. E. (1990). *Audience and rhetoric: An archaeological composition of the discourse community.* Englewood Cliffs, NJ: Prentice Hall.

Rehling, L. (1999). Print to online: Conflicting tales of transition. *Technical Communication, 46,* 27–35.

Rockley, A. (2002). *Managing enterprise content: A unified content strategy.* Indianapolis, IN: New Riders.

Rogers, E. M. (1962). *Diffusion of innovations* (1st ed.). New York: Free Press.

Rogers, E. M. (2003). *Diffusion of innovations* (5th ed.). New York: Free Press.

Russell, D. (1995). Activity theory and its implications for writing instruction. In J. Petraglia (Ed.), *Reconceiving writing, rethinking writing instruction* (pp. 51–77). Mahwah, NJ: Lawrence Erlbaum Associates, Inc.

Russell, D. (1997). Rethinking genre in school and society: An activity theory analysis. *Written Communication, 14,* 504–554.

Russell, D. (1998). Writing and genre in higher education and workplaces: A review of studies that use cultural–historical activity theory. *Mind, Culture, and Activity, 4,* 224–237.

Sharif, N. (2005). Contributions from the sociology of technology to the study of innovation systems. *Knowledge, Technology, and Policy, 17*(3–4), 83–105.

Spinuzzi, C. (1996). Pseudotransactionality, activity theory, and professional writing instruction. *Technical Communication Quarterly, 5,* 295–307.

Spinuzzi, C. (1999a). *Designing for lifeworlds: Genre and activity in information system design and evaluation.* Unpublished doctoral dissertation, Iowa State University, Ames.

Spinuzzi, C. (1999b). Grappling with distributed usability: A cultural–historical examination of documentation genres over four decades. In J. Johnson-Eilola & S. Selber (Eds.), *ACM SIGDOC '99* (pp. 16–21). New York: ACM.

Spinuzzi, C. (2002). Compound mediation in software development: Using genre ecologies to study textual artifacts. In C. Bazerman & D. Russell (Eds.), *Writing selves/writing societies: Research from activity perspectives* (pp. 97–124). Retrieved October 3, 2005, from the WAC Clearinghouse website: http://wac.colostate.edu/books/selves_societies/

Sullivan, P., & Dautermann, J. (1996). *Electronic literacies in the workplace: Technologies of writing.* Urbana, IL: NCTE and Computers in Composition.

Swanson, E. B., & Ramiller, N. C. (1997). The organizing vision in information systems innovation. *Organization Science, 8,* 458–474.

Swanson, E. B., & Ramiller, N. C. (2004). Innovating mindfully with information technology. *Management Information Systems Quarterly, 28,* 553–584.

Trotter, P. (2005, June 10). Re: Intro for students interested in single-sourcing? [stc-single-sourcing-l]. Message posted to http://lists.stc.org/cgi-bin/lyris.pl?enter=stc-single-sourcing-l

Venkatesh, V., Morris, M. G., Davis, G. B., & Davis, F. D. (2003). User acceptance of information technology: Toward a unified view. *Management Information Systems Quarterly, 27,* 425–478.

Walsham, G. (1995). The emergence of interpretivism in IS research. *Information Systems Research, 6,* 376–394.

Wegner, D. (2004). The collaborative construction of a management report in a municipal community of practice: Text and context, genre and learning. *Journal of Business and Technical Communication, 18,* 411–452.

Weick, K. E. (1995). *Sensemaking in organizations.* Newbury Park, CA: Sage.

Winsor, D. (1999). Genre and activity systems: The role of documentation in maintaining and changing engineering activity systems. *Written Communication, 16,* 200–224.

Winsor, D. (2000). Ordering work: Blue-collar literacy and the political nature of genre. *Written Communication, 17,* 155–184.

Yates, J., & Orlikowski, W. (1992). Genres of organizational communication: A structurational approach to studying communication and media. *Academy of Management Review, 17,* 299–326.

David Dayton is an assistant professor in the English Department at Towson University (Maryland). He teaches undergraduate courses in technical communication and graduate courses in technical and scientific communication, digital information design and architecture, and user-research methods.

Intercultural Rhetoric, Technology Transfer, and Writing in U.S.–Mexico Border Maquilas

Barry Thatcher
New Mexico State University

This article explores the transfer of U.S. technologies to three *maquilas*, or joint U.S.–Mexican manufacturing facilities in northern Mexico. Drawing on case study methods, it focuses on the rhetorical strategies that Mexican engineers and manufacturing personnel used to translate U.S. technologies and corresponding documentation for their Mexican contexts. It also suggests ways U.S. technical communicators can adapt their documentation to be more effective for these U.S.–Mexican intercultural rhetorical contexts.

Many studies have attempted to explore the integration of new technologies in specific contexts (Hawisher & Selfe, 2000; Munir, 2002; Orlikowski, 1992; Zuboff, 1989). This research has appropriately complicated the political, social, cultural, and contextual natures of new technology integration (Feenberg, 1996), particularly in rhetoric and writing technologies (Hawisher & Selfe, 2000; Sullivan & Porter, 1997). Many theorists argue that technology integration is a co-construction between raw technologies, which restrain and reinforce certain social and technological possibilities, and the cultural context of corresponding human actors (Feenberg, 1996; Orlikowski, 1992; Zuboff, 1989).

This co-constructive relationship sheds light on technology transfer in numerous contexts. However, in international settings, the relationship becomes more complicated because of the distinct actors, relationships, and contexts. In these co-constructive relationships, technologies do not relate to or fit each cultural and rhetorical tradition in the same way (Trompenaars & Hamden-Turner, 1998). Rather, these technologies develop complex relationships to each cultural–rhetorical tradition across the globe. This relationship evolves as the cultures and technologies evolve (Thatcher, 2005). Thus, the relationship between technology and culture is one of fit and reciprocity. *Fit* is the degree of correlation between the technology and the culture, and *reciprocity* is how fit evolves as both the technology and culture evolve and mutually influence each other (see Thatcher, 2005).

Fleshing out the relationship between fit and reciprocity has been difficult for both scholars of international technology transfer and U.S. scholars of rhetoric and technology studies. In international technology transfer, scholars have traditionally ignored the importance of culture, not to mention intercultural communication issues (see, for example, Bozeman, 2000). However, a few scholars have begun to research and document the influence of culture in integrating new technologies (see Bhagat, Kedia, Harveston, & Triandis, 2002; Munir, 2002). Munir (2002) explained how the long tradition of ignoring culture has impoverished research in international technology transfer and argued for understanding culture's role. However, the predominant approach in this newly emerging emphasis in international technology transfer studies has reflected an incomplete view of cultural influences on technology transfer. Munir reduced the influence of culture to the organization's cognitive and normative institutions (p. 1411). The term *normative* seems to represent the institutional roles and personalities of those who fill those roles. The term *cognitive* seems to derive from the larger cultural heritage, which influences decision making at the organization, but Munir explicitly argued that these larger cultural patterns tend to be deterministic and cannot account for the complex interplay at the organizational level (p. 1407). Munir's local approach was unsatisfactory because he failed to understand the roles of broader cultural and rhetorical patterns in daily sensemaking activities, and thus he could not account for the range and extent of influence of culture on organizational behavior (see Trompenaars & Hamden-Turner, 1998).

Like Munir (2002), many U.S. scholars in rhetoric and technology have favored the localized approach in their studies of both U.S. and intercultural contexts. Furthermore, some U.S. scholars have framed their understanding of fit and reciprocity on U.S. rhetorical and cultural traditions that reflect an essentialized and often ethnocentric approach. This American approach does not apply to other cultural and rhetorical traditions (see Thatcher, 2005; Trompenaars & Hamden-Turner, 1998). For example, U.S. theories that emphasize the local approach to technology–rhetoric relations (Hawisher & Selfe, 2000; Sullivan & Porter, 1996) correlate strongly with U.S. values of individualism, common law legal traditions, and universalism. Not surprisingly, this correlation presupposes a certain rhetorical purpose for technical communication as a medium for technology transfer. This purpose emphasizes a universalizing and precedent-setting function of writing in U.S. rhetorical traditions. However, predominant Mexican rhetorical traditions, for example, are based more on interpersonal orientations, civil law legal traditions, and particularism. Thus, these traditions create different assumptions about the rhetorical purpose of technical communication and technology transfer. Technical writing in Mexico has a strong notary-like purpose, serving to authorize uses of technology rather than guide its use, which occurs through interpersonal and particular relations. A number of recent studies have indicated that valid inquiries into intercultural communication have to account for the broad rhetorical and cultural patterns that originate in larger cultural–historical contexts but serve as a repertoire of strategies writers appropriate and individualize to make sense of everyday rhetorical situations (Thatcher, 2001). These broader rhetorical patterns often become standardized in a particular way in an organizational culture, becoming genres of professional communication based on the organization or profession. Further, people subject these professional and organizational patterns to serve their own purposes, constantly revising them. Therefore, intercultural researchers need to assess how personal appropriations of the strategies influence the development of these broad or organizational patterns (see Thatcher, 2000, 2001). Most likely, an organization transferring new technology would draw on four levels of cultural and rhetorical patterns to integrate that technology (Thatcher, 2000): (a) the larger cultural context, which provides general attitudes about the technology; (b) the local or regional context; (c) the specific organizational culture of the site; and (d) the personalities of those within the organization who draw upon the previous three patterns to make sense of their everyday activities.

Critical questions then for this study focused on the relationship among rhetoric, culture, and purposes of technical communication for technology transfer.[1] How much do the purposes of technical communication in technology transfer reflect the rhetorical tradition from which the technical communications come? What kinds of adaptations are needed in technical communications to meet the rhetorical needs of other cultures in a technology transfer situation? How do the larger cultural patterns encourage local, organizational, or personal rhetorical strategies when integrating new technologies? Finally, what would a cross-cultural examination of rhetoric–technology relationships tell us about what has been naturalized (Stewart & Bennett, 1991) with U.S. approaches to these very relationships?

To answer these questions, this article examines the roles of technical communication in the transfer of new technologies in four U.S.-owned manufacturing plants or *maquilas* in Mexico. A cross-cultural inquiry unmasks assumptions from both cultures about the roles of writing in creating or maintaining certain types of rhetorical, social, organizational, and cultural interactions in new technology contexts. This article first explores the major rhetorical traditions of Mexico and the United States, demonstrating why rhetoric and technologies have different relationships. It then explains the 4 × 4 qualitative research design, and finally, it explores the results.

RHETORIC AND TECHNOLOGY IN MEXICO
AND THE UNITED STATES

This section historicizes the use of technical writing in Mexico and the United States, especially as it is relates to science and technology.

---

[1]The research reported in this article was approved by the institutional review board for human subjects research at my institution.

History of Rhetorical Traditions

Mexico and the United States were settled at about the same time: Mexico in 1521 with the fall of Tenochtitlan, and the United Sates in 1607 with the founding of Jamestown. Although both areas attracted new colonizers, the motivations of the colonizers and the ensuing social and cultural orders that evolved "could not be more different" (LaRosa & Mora, 1999, p. 1). The Spanish colonizers were motivated by economic and religious purposes, and they constructed their societies based on "Iberian institutions and priorities" (La Rosa & Mora, 1999, p. 1). The most important cultural and social mechanism of the Iberian institution was the *encomienda*, or labor grant. A Spanish soldier or colonist was granted a certain track of land or village together with its native inhabitants. The Spaniards routinely set up a strict hierarchy over the land and its inhabitants, effectively creating an intricate caste system. An important part of this intricate caste system was intermarrying between races and social classes, effectively creating innumerable levels of social hierarchy. The encomienda also allowed for the establishment of the Catholic Church, which tended to dominate education and social life. Many of the Spaniards amassed great fortunes and were able to return to Spain as nobleman (La Rosa & Mora, 1999).

The United States had a remarkably different historical formation and corresponding social, cultural, and political organization. Many of the original colonies were settled by Dutch and English religious dissidents escaping religious persecution in Europe and had little desire to return there. Unlike the colonizers in Latin America, Dutch and English religious dissidents were automatically skeptical of authority, especially the combination of traditional religious authority and state government. These settlers established small, highly independent communities of "like-minded individuals" (La Rosa & Mora, 1999, p. 2). The U.S. colonizers did not intermarry with the indigenous populations; they just exterminated them or forced them onto reservations. The U.S. colonizers had much weaker ties to their mother countries than did their Latin American counterparts. U.S. independence was much shorter and easier to carry out, resulting in breaking with the cultural, religious, and social order of the mother country; the break was relatively easy and abrupt. In contrast, the Latin American journey to independence was much more ambiguous and complex, and most important, the independence gained by the Latin American countries did little to change the cultural, religious, and social order that existed in the pre-independence society. Latin American independence has been characterized as " same mule, different rider" (La Rosa & Mora, 1999, p. 3).

These two different foundations eventually yielded strikingly contrastive cultural and rhetorical traditions. In Latin America, the result of the encomienda foundation is a collective, yet highly stratified, culture where personal identity, world view, and interpersonal relations are based on the individual's kinship or social group. According to many intercultural theorists, Mexico consistently ranks high in collective values (see Hofstede, 1997; Trompenaars & Hamden-Turner, 1998). Osland, de Franco, and Osland (1999) argued that the high collectivity in this region differs from the horizontal, often harmonious, collectivity common in other parts of the world, such as Japan. Latin American collectivity is very hierarchized, creating strict observance of in- and out-groups.

These cultural and rhetorical differences also correlate strongly with organizational management approaches. According to many theorists, management in Mexico valorizes dependency, hierarchy, and close overseeing of subordinates' activities, whereas management in the United States tends to emphasize more independence and equality, leaving a reasoned amount of problem solving to the subordinates (Hofstede, 1997; Kras, 1989; Trompenaars & Hamden-Turner, 1998). Closely related to this distinction is the concept of *power distance*, which measures the ability of two people with different power and authority to influence each other (Hofstede, 1997). In Hofstede's (1997) measure, among all countries researched, Mexico had the second highest power-distance score: The subordinate has very little influence on the superior, but the superior has significant influence on the subordinate. The U.S. generally has much lower scores, indicating more mutual influence. Thus, both manager and subordinate in Mexico prefer that the superior closely oversee the work of the subordinate, whereas the U.S. style is more consultative. In Mexico, strict adherence to the communication patterns associated with high power-distance culture is common. These patterns reflect authoritative relationships that are often set up in the home, church, and school and then are carried to the workplace (Kras, 1989). Communication in lower power-distance cultures is often more consultative, involving more feedback, creativity, and flexibility in interpretation and application.

Relations of Rhetorical Traditions to Technologies and Writing

These two cultural systems exemplify and reinforce different rhetorical traditions that significantly influence the teaching and integration of new technologies. First, when colonizing Mexico, the Spaniards conveniently drew upon the rhetorical traditions of the native populations (Maya, Aztec, Mixtec, and Zapotec) for their colonizing purposes (see Marcus, 1992). Marcus (1992) explained that all four indigenous populations themselves used writing as a colonizing tool: They rewrote history for the benefit of their group, developed propaganda in their own interest, instilled myths appropriate to their positions of power, and established class markers. Thus, when the Spaniards arrived in Mexico, it was easy to use writing and written discourse the same way, and many researchers have argued that they did (see León-Portilla, 1996). For example, Kellog (1995) explained that the Spanish used the legal system and legal documents "as a powerful tool of acculturation, profoundly altering Mexico and Nahua conceptions of family, property, and gender. And it played a critical role in establishing and maintaining Spanish cultural hegemony" (p. xxix). As a result, writing became associated with all the elements of colonialization, serving what Kellog called notary-like functions.

This rhetorical tradition of writing as a colonizing and notary-like mechanism contrasts remarkably with U.S. traditions that valorize constructive and precedence-setting functions such as the signing of the U.S. Declaration of Independence. According to many comparative legal scholars (Alcalde, 1991; Rosenn, 1988), the signers of the Declaration of Independence and the U.S. Constitution were able to rely on a cultural and rhetorical context that enabled such a use of written communication. This is why these documents have worked so well in the United States, but comparable documents have had significantly less influence on Latin American countries (Alcalde, 1991). In the United States, writing reflected and reinforced traditions of individuality, universalism, equality, and common law reasoning (Thatcher, 2000). On the other hand, the Mexican rhetorical traditions reflected and reinforced an in- and out-group orientation that is common in collective cultures, hierarchical social organizations, and particular or relational thinking patterns (Thatcher, 2000).

Consequently, new technologies meet different cultural and rhetorical assumptions when they are deployed in Mexico and the United States. In Mexico, there is probably a much stronger association of the new technology with colonialism and hegemonic cultural and economic structures, especially when that technology comes from a colonial power such as the United States. This association of colonialism and hegemony is compounded when the new technology is deployed with written documentation, another traditional tool of hegemony.

Maquila Situation

The transfer of new technologies to Mexico also needs to be situated in U.S.–Mexican history of relations and the history of the maquila (see Garza-Almanza, 1999). Maquilas are manufacturing plants located in Mexican cities bordering the United States. During World War II, when almost all available U.S. labor was used for war-associated manufacturing, the U.S. government agreed with Mexico to allow Mexican nationals to work in the United States legally in nonwar-related areas such as agriculture. Known as *braceros*, as many as 5 million Mexicans worked in the United States during this time (Garza-Almanza, 1999). Many braceros stayed, but under the agreement, many returned to Mexico. In 1961, the United States abruptly ended this agreement, leaving many Mexicans either stranded in the United States or in northern Mexican border cities.

The large number of stranded Mexicans in northern border cities prompted the Mexican government to study the feasibility of a free trade zone to take advantage of these cities' proximity to the United States and a surplus of inexpensive labor. The result of this research led to the development of the manufacturing industry of exportation, better known as *maquiladora*, which began in 1966. The maquila industry has flourished since its beginning. There are more than 3,800 manufacturing plants employing more than 500,000 people along the U.S.–Mexico border (Garza-Almanza, 1999). Materials are imported into Mexico where low-wage maquila workers typically perform the assembly portion of the manufacturing process. The finished products are then returned to the United States or other countries.

Research Questions: Technology Transfer in an Intercultural, Border Context

Because maquilas have more than 40 years of history in Mexico, many border areas are accustomed to U.S. rhetorical and cultural patterns. As detailed later, almost 90% of the maquilas are U.S.-owned and many are U.S.-managed. Crucial questions for border research focus on the hybridity of the maquila. What kinds of rhetorical traditions exist in the maquila situation? What elements of Mexican and U.S. rhetorical traditions thrive in this border environment? What elements are lost? How does this border situation influence the transfer of new technologies and what documentation and training strategies best serve this unique intercultural situation? Finally, how do these results reshape U.S. theory and practice of rhetoric and technology relationships?

RESEARCH DESIGN

Qualitative Approach and Multiple Case Studies

To answer these questions, this study used a qualitative research design that was semi-exploratory, capable of assessing and designing communication and training materials in a cross-cultural situation. A multiple case study design is best suited for this context (Thatcher, 2000; Yin, 1994). First, because there was existing knowledge of maquila cultures, the study did not need the exploratory nature of an ethnography of a single site. Also, because the study sought to develop methods and materials that were applicable beyond one site, the study needed to assess training in new technologies at more than one site. Thus, a focused exploration of how multiple organizations in the border area have effectively developed new training systems for new technologies was the most appropriate design. Because the research design needed to be qualitative to assess cultural and communicative training needs, the number of research sites had to be limited. As detailed in other intercultural research on training in new policies and procedures (Thatcher, 2000), four sites provided an effective balance between the need for generalization and the complex management of qualitative data-gathering and analysis.

The first research site was a U.S.-owned maquila in Chihuahua City, Mexico (250 miles south of El Paso, Texas). This maquila manufactured cruise controls for a major U.S. automobile manufacturer. The managers invited me to research their training processes because they hoped to improve these processes to obtain International Organization for Standardization (ISO) certification. Although this maquila was U.S.-owned, all its management and personnel were Mexicans, mostly from northern Mexico. This maquila had about 200 employees at the time of the study, making it a small to mid-sized maquila. It had been in operation since 1991. The subject of the case study for this site was the communication processes used to train line personnel in new manufacturing procedures.

The focus of study for sites 2 through 4 was the training requirements for ISO 14000 Environmental Certifications for these Juárez, Mexico, maquilas. This research was funded by an Environmental Protection Agency (EPA) grant, and I collaborated with a Mexican university professor. Site 2 of our study was a U.S.-owned maquila in Ciudad Juárez, which borders El Paso, Texas. Site 2 built complex electronic sensors for U.S. armed forces headquarters in the northeastern United States. This site also had all Mexican management and personnel, about 500 employees. Site 3 was a U.S.-owned maquila in Ciudad Juárez that built smoke detectors for the U.S. and global market. This site was a small maquila of about 100 employees with all Mexican management. Site 4 was a large international U.S.- and Dutch-owned conglomerate where consumer electronics—mostly televisions—were manufactured. This maquila was large, with about 4,000 employees. All managers and personnel were Mexican.

The units of analysis for all four cases focused on the relationships of U.S. and Mexican rhetorical traditions, border maquila situations, and training in new technologies.

Data Gathering and Analysis

Data gathering occurred in two phases. The first phase, completed in less than a year, exclusively focused on the training patterns of the first maquila. The next phase, which lasted two years, tested the series of hypotheses and theories (Flick, 2002; Yin, 1994) developed from the first site. At all four sites, this study utilized a variety of data-collection methods, which is the strength of the case study design (Flick, 2002; Yin, 1994). The data collection followed three basic stages common to qualitative research: ethnographic exploration, examination of the four sites, and the poststudy.

Stage 1.   At all four sites, we used five ethnographic explorations of the culture and discourse of the two organizations: (a) analysis of the physical and social setting to understand broad communication- and document-cycling patterns; (b) analysis of managerial procedures, policy, and company information to see how the corporate culture fit the contrasting cultural and social values; (c) formal and informal interviews about the corporate culture and discourse patterns; (d) exploratory observations of social interactions and management procedures; and (e) collection of, and interviews about, previous discourse to note genres and discourse conventions.

Before we began formal data gathering of sites 2 through 4, we held a roundtable meeting with 15 environmental engineers from major maquilas in Juárez. The meeting was held at the Autonomous University of Ciudad Juárez (UACJ). In the meeting, a UACJ master of science student and researcher for this grant guided the group through a 21-question discussion of ISO 14000 implementation in Juárez maquilas. This student and my coprincipal investigator for the study, Victoriano Garza-Almanza (UACJ professor of environmental engineering), created this diagnostic questionnaire as a way to assess engineers' level of ISO 14000 implementation.

Stage 2.   The second stage used six interactionist and ethnomethodological designs to examine the interactions of the four cases (Thatcher, 2001): (a) collection of all key documents in the discourse project; (b) observations and logs of social interactions such as formal and informal meetings, negotiations, and reviews among authors/writers, reviewers, and readers; (c) discourse analysis of all key documents (Flick, 2002); (d) process- and discourse-based interviews of authors/writers, reviewers, and readers about their key processes and products of the discourse project; (e) discourse-based preference surveys asking informants to choose between different rhetorics (Thatcher, 1999); and (f) qualitative and quantitative surveys. I used thematic coding to study sites 2 through 4 (Flick, 2002) and evaluated the data in light of what I observed in site 1, researched in South America (Thatcher, 1999), and read in existing literature. I also used theoretical coding, or grounded theory (Flick, 2002), to build a new and dynamic theoretical framework for the data.

Stage 3.   The third stage was poststudy research to check my findings with the participants, verify my findings, and explore the typicality of this research situation. I carried out retrospective, formal, and informal interviews of authors/writers, audiences, and reviewers.

RESEARCH RESULTS

The results of this study show that integrating new technologies in U.S.-owned but Mexican-managed maquilas requires addressing the complex border situation, which comprises elements of both Mexican and U.S. rhetorical and cultural traditions. The results also show that U.S. researchers and technical communicators cannot assume the same relation of technology and documentation in other cultures as they find in their own culture. Researchers also need to denaturalize their assumptions about these relationships to understand them even better in U.S. cultural contexts. This results section first narrates the complete picture of technology transfer in the Chihuahuan maquila, which serves as a nice introduction to specific issues that will be explored in the subsequent four sections.

Technology Transfer in Chihuahua Maquila

The Chihuahuan maquila site, which was put into operation in 1991, allowed me to interview those who were still familiar with integrating the new technologies sent from the home factory in Ohio. The head manufacturing engineer, Eduardo (pseudonym), related the process of this deployment during our interview. He explained that the home office in Ohio sent an American to Chihuahua as a head engineer to oversee plant development; four additional Americans were sent throughout the deployment process. The head engineer, John McGregor (pseudonym), remained four years in Chihuahua. He received specific instructions from the home office in Ohio and was responsible for relaying those instructions to the Mexican engineers. During the first four years, approximately 15 Mexican engineers were hired to get the plant up and running.

The following transcription of our interview with Eduardo explains that the critical role of the Mexican engineers in this integration process was that of translating the new technology for the Mexican line workers. My English translation is followed by the Spanish original:

> Barry: Tell me about the development of the maquila and your work with those from the United States. [*Cuéntame del desarrollo de la planta y del trabajo con los de los estados unidos.*]
>
> Eduardo: Ah, when it was decided to start up the plant here, uh, the first thing that was done after building the building was to bring here three or four engineers. One of them [pause] ... was the plant manager for various years, uh, the engineering manager, Mr. John McGregor. He was the one that gave instructions about how the machinery should be setup. [*Eh, cuando se decidió que se arrancara aquí la planta, eh, lo primero que hicieron la parte después de construir el edificio fue traerse aproximadamente tres o cuatro ingenieros, uno de ellos, eh, le emplearon número uno que fue gerente de planta durante varios años; ah, un gerente de ingeniería, el señor John McGregor, que fue el que dio instrucciones de cómo debía hacer la maquinaria.*]
>
> Barry: U hum.
>
> Eduardo: Ah, what was done at that time was that the engineering personnel from the United States in collaboration with John McGregor saw

what had to be done and told the contracted engineers. [*Ah, lo que hacían en aquel entonces era el personal de ingeniería de estados unidos en colaboración con el señor Joe Gregory veían que era lo que se tenían que hacer y les decían a los ingenieros que han sido contratados.*]

Barry: In Chihuahua? [*¿en Chihuahua?*]

Eduardo: In Chihuahua—what it was that they had to do; and they were in charge of telling the operators [line workers] how to carry out their operations. [*En Chihuahua—que es lo que se tenía que hacer; y ya ellos encargaban de decirles a los operadores cómo realizar las operaciones.*]

Barry: Thus, since the beginning, the engineers have been the translators, so to speak. [*Entonces, desde el principio los ingenieros han sido los traductores, digamos.*]

Eduardo: That, exactly. [*Eso, exactamente.*] (B. Thatcher, personal communication, September 5, 2003)

Later, I asked Eduardo about why the engineers had to be the translators of the technology transfer, how they ended up doing that job, and how they decided upon the best means of technology integration. Eduardo said that the original U.S. "instructions were very vague" and that the Mexican engineers could not figure out what to do with them. Finally, as their frustration grew, so did the necessity to understand fully the operation of the equipment. So they ignored the manuals and tried to understand the machinery in their own way:

Eduardo: Once we got rid of our fear and begin to see how they functioned, we begin to take them apart and begin to understand all of the black magic. Thus, this was when we were finally able to really improve many, many, many processes. It was like EVERYTHING depended on ALL the instructions given by the American engineers, and the people in the United States did not touch anything. Yes, faithfully, we followed the orders until the moment arrived that we begin to have initiative; thus, we begin to see the problems and begin to propose solutions. [*Y una vez que nos quitamos el miedo con trabajar con las máquinas, este, empezamos ya a ver cómo funcionaban, empezamos a desarmarlas y empezamos a entender todo esa magia negra. Entonces, fue cuando ya realmente pudimos mejorar muchos, muchos, muchos procesos. Este ... en un principio, es como TODO depende de TODAS las instrucciones que tenían dados por los ingenieros americanos; y gente que está en estados unidos no se tocaba nada. Sí fielmente seguíamos las órdenes hasta que llegó un momento en la que empezó a tener iniciativa, entonces, empezaban a*]

> *ver los problemas y se empezaban a proponer soluciones.*] (B.
> Thatcher, personal communication, September 5, 2003)

After understanding the machinery from their own perspective, Mexican manufacturing engineers were able to improve several processes ("many, many, many processes"). Later, Eduardo detailed how the Mexican manufacturing engineers, armed with their own experience with the machinery, wrote their own machinery specifications that eventually became the basis for the written documentation of the work processes at this maquila.

As the next four sections explain, this translation process at the four maquilas had four components: (a) situating the roles of written documentation in Mexican organizational culture; (b) understanding the roles of documentation, hierarchy, and information control; (c) understanding the high turnover rate as it related to problem solving and documentation requirements; and (d) understanding the profiles of technical communicators and documentation.

Writing and Technology in Mexican Organizational Culture

One major obstacle in U.S.–Mexican technology transfer is the different expectations in management styles. Mexico's high power-distance modes of management correlate strongly with linear technology transfer; that is, superiors' instructions are followed with little questioning, including those for the integration and uses of technology. Conversely, U.S. managers assume that subordinates will ask many questions and function in a partner-like manner during technology transfer.

In the Chihuahuan maquila, Eduardo and the other Mexican engineers were working in the high power-distance model until they saw it failing; the Mexican engineers were afraid to disregard the specific instructions from the American engineers. However, the Mexican engineers' initiative was not punished, and when they exhibited initiative, the plant flourished. It was impossible to interview the American plant manager who had left 6 years earlier, but it seems possible—based on my experience in similar U.S.–Latin American interactions (Thatcher, 1999)—that the U.S. engineers expected the Mexicans to take the initiative and learn the engineering processes for themselves. However, the Mexican engineers who were used to high power-distance management were nervous about doing so.

Documentation, Hierarchy, and Information Control

This high power-distance management mode resituates the roles of documentation and information control in the organizational hierarchy, becoming almost a complete reversal of many U.S. approaches to technical communication. Different assumptions about power found in the Chihuahuan maquila similarly played an important role in the training for ISO 14000 certifications at sites 2 through 4. The initial group interview was structured around a 21-question assessment of the different parts of the ISO 14000 process, including the role of documentation and manuals. Each of the 21 questions had a range of possible of answers ranked from 1 (indicating weakest role for documentation) through 5 (indicating strongest role for documentation). The following question assessed the roles of manuals and documentation in the environmental management processes (my English translation):

Concerning formal environmental documents and procedures manuals, the maquila

1. Has nothing.
2. Local administration prepares manuals and documentation when necessary, sends copies, and whoever needs them takes a copy.
3. Manuals and documentation are available for the majority of activities that have significant impact on the environment, and copies are sent and are available for those who need them.
4. Manuals and documentation are comprehensive, referencing all activities with significant environmental impact and are distributed according to agreed upon circulation lists.
5. Manuals and documentation are comprehensive and are revised periodically; their distribution is authorized by proper authorities, are distributed according to agreed upon circulation lists, are available in all facilities where they are needed, and are removed when obsolete.

The average score given by the 15 environmental engineers representing a range of maquilas in Juárez was 2.6, a weak score. More important, these questions invoked a strong debate (*polémica*) in the group interview about the roles and types of documentation needed in the organization. Some environmental engineers did not want to document their work procedures because they believed oral training was best. However, oral training had many drawbacks, including the problems of hierarchical management systems and the need for constant retraining due to high turnover rates. Some environmental engineers also understood that reliance on oral training fostered both an inability to solve problems and a shirking of responsibilities. Other engineers maintained that traditional lines of authority and decision making needed to be in place to effectively manage personnel. A few engineers, to maintain their control of the manufacturing processes, strongly asserted their right to limit the amount of documentation.

Garza-Almanza, my Mexican colleague in this EPA-sponsored study, had written the 21-question assessment, and, knowing my approach to the problem, listened to the debate and then asked my opinion. I told him that I believed that the issues of high turnover rates and low problem-solving skills (to be discussed later) could be addressed by better written documentation. Two engineers vehemently disagreed with me, never linking problem solving and effective training to written documentation. From their perspective, the best training was to have the on-site engineer who was present oversee all functions at all times. They feared that written documentation would bring line-worker independence, which could introduce errors into the manufacturing process.

Another maquila environmental engineer who was a medical doctor served as the health and environmental specialist for the smoke-detector maquila. She immediately saw the link between high power-distance modes of management and problem-solving skills at her maquila. She quickly challenged the two engineers about their high power-distance methods. Garza-Almanza then explored his general frustration that hierarchical management styles promote passivity in Mexican line workers, a passivity that he sees in Mexican culture in general, regarding the integration of new technologies. From his perspective, there is great resistance to empowering employees to solve problems or influence policy, procedures, or workplace practices. This high power-distance relationship was not isolated to this group interview. Lines of authority in all four maquilas were very clear. Rarely were subordinates in positions to influence training.

Turnover, Problem Solving, and Documentation

This high-power organizational culture combined with a high turnover rate for employees worked against a culture of effective training. Managers or superiors at all four maquilas insisted on maintaining control through interpersonal power relations, a type of training method that takes much time to develop so that it functions well (see Victor, 1992). However, the maquila industry is consistently plagued by high turnover rates. In good economic times with plenty of job opportunities, many employees job hop—that is, move from job to job looking for the best wages and situations. In less prosperous times, on the other hand, limited training budgets become an issue. For example, during the course of this study (August 2003–December 2004), a large number of maquilas laid off employees and moved their operations to Asia, where production costs are lower. As a result, most maquilas had even tighter than normal operating budgets. Personnel at all four maquilas noted that training was the first budget item to be eliminated. In the large Dutch maquila, almost all surveyed employees expressed a desire and need for more training.

As Victor (1992) explained, management in high power-distance and high-context cultures require a capable person at the top to orchestrate all relationships. When this person is capable and functions as expected, the organization runs well. However, there is considerable lag time when personnel changes occur (pp. 138–158). This finding correlates with the four maquilas we studied. Despite the desire of some engineers to tightly control operations on an oral basis, many Mexican personnel saw a need to change the organizational culture through more written communication. This desire was sparked by what Mexican management perceived as a lack of initiative and problem solving at the line-worker level. Eduardo, the head manufacturing engineer at the Chihuahuan maquila, linked effective written documentation to initiative and problem solving. When I asked Eduardo what kind of documentation he would recommend that the home base in the U.S. send to a maquila in Mexico, he answered

> First, the most basic, that those responsible understood first how things work, the machinery as well as the product—so that they would be able to base their information on what to do. One thing we had here when the plant started was the contracted people that normally felt uneasy about working and taking initiative … if it's something new, with initiative to responsibility [is needed] because initiative without responsibility causes, can create many, many problems that can be worse. What's better—the key word—the WHY of the things. [*Lo básico primero sería, este, que los responsables entendiesen primero cómo funcionan las cosas, tanto la máquina como el producto—ya para poderlos basar, para hacer información sobre qué hacer. Una cosa que tuvimos aquí cuando inició la planta era la gente que era contratada, era normalmente gente que sentía inquietud para trabajar y tener iniciativa … si es algo o un proceso totalmente nuevo, con iniciativa a responsabilidad, porque iniciativa sin responsabilidad causa, se pueden crear muchísimos problemas que puedan ser peores … Eso mejor—es la palabra clave—el PORQUE de las cosas.*] (B. Thatcher, personal communication, September 5, 2003)

According to Eduardo, the strict oral and hierarchical management mode produced workers that did not have initiative or responsibility. Thus, they were not valued as much as those who used combined initiative and responsibility. When I asked Eduardo how workers can combine initiative and responsibility to understand maquila operations, he commented that fostering problem solving based on good documentation was key. In an interesting move, he argued that this combination of initiative, responsibility, and problem solving was precisely what the ISO 9000 certifications produced when their maquila obtained it a few years earlier.

From a similar perspective, in the Dutch maquila, line workers were surveyed about the kind of training they liked most: oral, written, or a combination both. Of 30 employees surveyed, only 3 preferred oral training; 13 preferred written training, and 14 preferred a combination. Many of those who argued for the written training wanted the "independence and time that written documentation affords when learning new procedures." None of the respondents linked written documentation to initiative, responsibility, and problem solving. This is not surprising given the lack of operational perspective exhibited by line workers. However, in addition to Eduardo (the head manufacturing engineer at the Chihuahuan maquila), the head plant manager at the Chihuahuan maquila, the doctor who was the health and environmental specialist at the smoke-detector maquila of site 2, and Garza-

Almanza all made the connections between well-written documentation and problem solving. They also argued for the necessity of doing it in Mexico.

One of the difficulties, however, in implementing written documentation in Mexico is the strongly correlated cultural values of interpersonal orientation, orality, writing, relatively low salaries, and the high power-distance modes of management. Perhaps the economic variable is most illuminating. Most environmental engineers at these maquilas made between $600 and $1,500 a month, a salary well above Mexico's minimum wage but well below U.S. averages. This is an important factor because developing effective documentation is actually more expensive than simply hiring another engineer to explain the processes that documentation would have explained in a U.S. context, for example. In addition, Mexico has been described as having a strong interpersonal orientation (Albert, 1996; Kras, 1989) associated with collective cultures and maintained by a strong preference for oral rather than written communication.[2] Written communication, as mentioned in the literature review, are, at times, strongly associated with official and legalistic communication purposes. Thus, these three variables (interpersonal orientation, dislike of writing, and greater cost of developing written documentation) strongly encourage an oral and high power-distance approach to training in new technologies. Consequently, using writing in Mexican maquilas as a medium of training faces significant obstacles that require, perhaps, key changes in predominant cultural values. Until there is a change in the predominating cultural values, obstacles that cannot be overcome via written communication will remain.

One force for change, however, seems to be the desire to obtain ISO certifications, which requires standardized and written documentations of work processes. As mentioned, the ISO 9000 Management Certification forced the Chihuahuan maquila managers to rethink their views of written documentation. And in the three Juárez maquilas undergoing ISO 14000 Environmental Certification that we studied, the need for written documentation seemed to encourage some openness and desire for change in traditional Mexican cultural and rhetorical traditions. It seems, however, that without these kinds of "outside forces," many Mexican maquilas will continue with high power-distance culture and oral methods to train personnel in new technologies.

Profiles of Technical Communicators in a Mexican Maquila

None of the four maquilas studied had a position labeled as a *technical communicator,* following what Garza-Almanza (2004) wrote: "In distinction to our country [Mexico], in the United States, science writing or scientific dissemination has obtained a social importance to the point that many universities have

---

[2]For a better explanation, see León-Portilla, 1996.

professionalized its duties. That is, have formally instituted the career path of a science writer" (p. 4; my translation). My work with other universities throughout Mexico confirms what Garza-Almanza said: Technical and scientific writing exists on a limited basis in Mexico for a variety of social, economic, and cultural reasons.

Although technical and scientific writing is uncommon in Mexico, there were written training materials in all four maquilas. Who was doing the technical writing? How was it being done? What was it like? In short, mostly engineers did the writing when they had the time and felt the necessity. Technical documentation development procedures at the Chihuahuan maquila was perhaps most revelatory in situating the profiles and roles of technical communication in Mexican maquilas. At this maquila, three employees did the technical writing, specifically, the procedure manuals for manufacturing cruise controls. The three writers were not called writers; rather, they were trainers in the human resource department, and as a way to improve their training effectiveness, they worked with the manufacturing engineers to create short manuals delineating line operations. All three trainers had much in common: They were young female workers from the surrounding area with no formal technical writing training. However, all three developed written technical documents to enhance their training programs. The profiles of the three writers are the following:

Ana Maria (pseudonym) had worked for this maquila for 4.5 years, 3 of which she functioned as an instructor/trainer and 2.5 years she wrote technical manuals. Like all the other personnel in the human resources department, Ana Maria began working as a line worker but was promoted to a technical trainer position because she demonstrated an ability to master the line operations quickly and was able to teach others effectively. Ana Maria, who was about 25 years of age at the time of the interview, was from Chihuahua City. She finished only her secondary (middle school) education, and she remembers receiving some instruction on orthography (grammar and syntax) but no formal writing instruction. She became a writer because she really liked to write.

Lourdes (pseudonym) was the second technical-manual writer at this maquila. She was born in a small pueblo in the state of Chihuahua, just south of Chihuahua City. Lourdes was forced to quit school after the fourth grade and work. She had worked at this maquila for 8 years, and before that she had worked at another U.S.-owned maquila in Chihuahua for 3 years. Like Ana Maria, Lourdes was also in her mid-20s. She was hired as one of the first line workers at this maquila and had been there "since the beginning," she said. She was also one of the first to create (*generar*) the job of writer/trainer at this maquila. She liked to write, and she enjoyed her job.

The third writer was Belen (pseudonym) who had recently graduated from a local university as an industrial engineer. She was in her mid-20s and had worked at this maquila for 6 years while completing her college degree. Belen was also from

Chihuahua City, and like Ana Maria, she was one of the first to develop her role as a technical writer. Belen was looking to continue her studies and obtain a master's degree in manufacturing engineering. She liked to write, but her heart was in engineering. She was also looking for a job as an engineer who designed and managed engineering manufacturing, rather than one as a trainer and writer.

Development and Roles of the Written Documentation

Like the profiles of these three writers, the roles of written documentation at this maquila reflect the cultural and rhetorical traditions of Mexico and maquila workers. When I asked the writers about the purposes of the manuals, they strongly argued for their great necessity and for the important functions that they filled. My conversation with Ana Mara is representative:

> Barry: What purpose do the manuals have? [¿*Qué propósito tienen los manuales?*]
>
> Ana Maria: What purpose? That the people know with what materials they are working so that each one works well with the parts with which they are working. This and analysis of materials, that they know how to handle the material. [¿*Qué propósito? Es que la gente sepa con qué materiales está trabajando para que sirva cada una de las partes con las que estamos trabajando. Este, y análisis de material, que lo sepan manejar el material.*]
>
> Barry: OK, and about training. Are the manuals an important part, of little importance, or very important? [*OK. En cuanto a la capacitación, ¿son los manuales una parte importante, poco importante, muy importante?*]
>
> Ana Maria: I believe they are a very important part because if they [workers] do not know something, they would be failing on the floor. And if they don't know what each one of the parts does, it's likely they do not know an adequate handling of the material. [*Yo creo que es una parte muy importante, porque, si no saben esto, estaría faltando ya en el piso. Y si no saben para que sirva cada una de las partes, y a lo mejor no saben el adecuado manejo de material.*] (B. Thatcher, personal communication, September 5, 2003)

Both Lourdes and Belen agreed with the importance Ana Maria placed on the manuals. However, Belen connected the use of written manuals to a theoretical—not practical and rote—understanding of the manufacturing process and accompanying problem-solving skills. Belen argued that the initial training and accompanying manuals were "provided for information transfer from one generation to another" ("*era de transmitir conocimiento en generación entre generación*").

However, the introduction of written manuals forced a focus on "criteria for judging engineering acceptability" and "reason" in the manufacturing process, something far beyond information transfer.

This difference between information transfer and criteria highlights the different approaches to manual development. Ana Maria and Lourdes essentially saw themselves in notary-like functions in the service of the engineers. They took the technical specification sheets from the engineers and put those sheets together, with minimal additions, to develop their manuals. However, every change—deletion or insertion—had to be certified by the engineers. Thus, it was not surprising that when I asked them if they liked being technical writers, they were a bit puzzled at the question. They did not consider themselves technical writers, but trainers/instructors who took engineering specifications and bound them in a usable format for the readers.

On the other hand, when I asked Belen if she liked writing manuals, she said

> Yes, I like to because it's a part of design. I don't know if you realized or saw, for example, that each person has her own method of designing the manuals, of structuring them. I like the design and development of manuals. [*Sí, me gusta porque es parte de diseño. No sé si usted logró captar o ver que por ejemplo cada área, cada persona tiene su manera de diseñar los manuales, de estructurarlos. Me gusta el diseño y la formación del manual.*] (B. Thatcher, personal communication, September 5, 2003)

In fact, I had noticed Ana Maria and Lourdes sensed that their manuals were different, but they never used the terms *design* or *development*. They just referred to their manuals as being different.

After commenting on that point—that Belen considered manual writing as a part of design—I asked Belen how she learned to do that kind of design. She laughed and said, "by myself" [*sóla*]. When I pressed her a bit more, she reasoned that perhaps because of her own gregarious and interpersonal skills, she was able to tell what audiences needed in the manuals and was able to design for them based on trial and error. Part of this ability was in visual rather than verbal thinking, for Belen had integrated the most visual elements in the manuals. In fact, I was surprised that her manual design followed many of the design principles that I teach at New Mexico State University in my technical writing classes using Anderson's (1999) *Technical Communication*. These principles include effective relationship of graphics and visuals, warning and troubleshooting, introduction, and theory of operation. I later presented Belen with a copy of this text, and she was delighted to see some theoretical discussion of what she had tried to do with her own technical manuals.

Ana Maria's and Lourdes's manuals were nearly all text and contained very few visuals. Most of their manuals were formatted according to the technical specifications of the engineers, not the manufacturing processes of the line workers. Belen had not linked her design approach to her bachelor of science degree in industrial engineering. But, when I pointed out the possible connection, she agreed that maybe there was a connection.

In light of these descriptions of developing procedures manuals, I was actually surprised about the roles that the manuals actually played in the manufacturing process. For workers to do a job on a specific line or post, they had to be certified to do that job. Certification required that they pass a written exam and then pass an on-site exam, demonstrating correct handling of the manufacturing process. When I heard about this certification process, I automatically assumed (from a typical U.S. individual approach) that the written manuals would be available to the employees to study by themselves at their own pace. However, I was wrong. In this training process, the trainers/instructors orally read the manuals to the trainees to familiarize them with the process. Then the instructors took the trainees to the line and demonstrated the process itself. The trainees then were given the technical specification sheets to study. Finally, the trainees took the written exam, which did not really test their knowledge of the procedures. Rather, the exam tested knowledge of rote technical specifications. The technical manuals were used only to introduce and—as I later hypothesized—to legitimize the process. The technical manuals were not an integral part of the hands-on learning that seemed so critical to the manufacturing process. To me, the manuals served much like the notary-like functions of writing that Kellog (1995) argued are still predominant in Mexico because of the country's legal heritage and institutions. When I asked Belen about this seemingly unproductive role of her manuals, she commented that the line workers did not want to use her manuals for anything more than an introduction to the general process of their line.

The surprising use of manuals in this Chihuahuan maquila brings the discussion full circle, in a way, when understood in the light of predominant Mexican rhetorical traditions. Even though Belen's manuals seemed entirely adequate for self-teaching of the process, which need not involve anybody else, these maquila personnel saw the manuals as fulfilling a very specific role of verifying and legitimizing a process that would be taught through oral and hierarchical methods. In this Chihuahuan maquila, I also surveyed a significant portion of the line workers, asking them what kinds of training media they preferred. Almost all preferred an oral medium rather than a written one. Thus, it was quite surprising to me to see in the large Dutch maquila an expressed desire by line workers for written documentation.

CONCLUSION: ANSWERING THE RESEARCH QUESTIONS AND RECOMMENDATIONS FOR FURTHER RESEARCH

The introduction and literature review set up the possibility of hybridity, a mixing of Mexican and U.S. cultural traditions in the Mexican-border maquilas. Did I see hybridity? Perhaps I did witness hybridity, but it was much less than what might be assumed by many border and postcolonial theories. Even though these maquilas were headquartered in the United States, which presumably brought its own corporate culture to Mexico, the traditional Mexican cultural and rhetorical traditions were clearly predominant in all four maquilas. More U.S.-like purposes for writing seemed to surface but did not take root, as exemplified by the notary-like functions of writing in the Chihuahuan maquilas. Thus, the border seemed to have little influence on the relations of rhetoric and technology, at least at these four maquilas.

Based on the little blending of rhetorical traditions, it is interesting to hypothesize about how much U.S. technical communication theory and practice is based on the assumption that people use writing to learn new technologies rather than to legitimize the technology so that oral methods of training can be effective. The lone technology user sitting at his or her desk with a pile of documentation trying to understand the technology is a U.S. vision of documentation, which probably rarely happens in Mexico. In addition to these findings, this study helps articulate some interesting questions about the relationship of technology transfer and rhetoric. It seems that written documentation was linked to problem solving and corresponding responsibility: a link that Ong and other scholars have argued for (see Kaufer & Carley, 1993, on this debate). Thus, despite the different purposes or uses of writing in the United States and Mexico, it seems as though users from both cultures agree with this one function of written documentation. Another important variable in this picture is the growing globalization of economies and corresponding drive for ISO certifications. These ISO certifications require written documentation of work processes. Thus, are these ISO requirements forcing a U.S. or European written rhetorical tradition on cultures that more comfortably work with more oral traditions? What might be the cultural ramifications of this written requirement?

REFERENCES

Albert, R. (1996). A framework and model for understanding Latin American and Latino/Hispanic cultural patterns. In D. Landis & R. Bhagat (Eds.), *Handbook of intercultural training* (2nd ed., pp. 27–48). Thousand Oaks, CA: Sage.

Alcalde, J. (1991). Differential impact of American political and economic institutions on Latin America. In K. W. Thompson (Ed.), *The U.S. Constitution and the constitutions of Latin America* (pp. 97–123). Lanham, MD: University Press of America.

Anderson, P. (1999). *Technical communication: A reader-centered approach* (4th ed.). Fort Worth, TX: Harcourt Brace.

Bhagat, R., Kedia, B., Harveston, P. D., & Triandis, H. (2002). Cultural variations in the cross border transfer of organizational knowledge: An integrative framework. *Academy of Management Review, 27,* 203–221.

Bozeman, B. (2000). Technology transfer and public policy: A review of research and theory. *Research Policy, 29,* 627–655.

Castañeda, J. G. (1995). *The Mexican shock: Its meaning for the U.S.* New York: New Press.

Feenberg, A. (1996). Subversive rationalization: Technology, power, and democracy. In A. Feenberg & A. Hannay (Eds.), *Technology and the politics of knowledge* (pp. 3–22). Bloomington: Indiana University Press.

Flick, U. (2002). *An introduction to qualitative research* (2nd ed.). London: Sage.

Garza-Almanza, V. (1999). Comercios regionales y ambiente: Integración del desarrollo y el ambiente en el tratado de libre comercio de América del Norte [Regional commerce and the environment: Integration of development and the environment in the North American Free Trade Agreement]. *Ambiente Sin Fronteras, 2,* 6–7.

Garza-Almanza, V. (2004). La divulgación de la ciencia en México [Dissemination of science in Mexico]. *Cultura científica y tecnológica, 1*(5), 3–16.

Hawisher, G., & Selfe, C. (Eds). (2000). *Global literacies and the World-Wide Web.* New York: Routledge.

Hofstede, G. (1997). *Cultures and organizations: Software of the mind.* New York: McGraw-Hill.

Kaufer, D., & Carley, K. (1993). *Communication at a distance: The influence of print on sociocultural organization and change.* Hillsdale, NJ: Lawrence Erlbaum Associates, Inc.

Kellog, S. (1995). *Law and the transformation of Aztec culture, 1500–1700.* Norman: University of Oklahoma Press.

Kras, E. (1989). *Management in two cultures.* Yarmouth, ME: Intercultural Press.

LaRosa, M., & Mora, F. (1999). Introduction: Contentious friends in the western hemisphere. In M. LaRosa & F. Mora (Eds.), *Neighborly adversaries: Reading in U.S.–Latin American relations* (pp. 2–18). Lanham, MD: Rowman &Littlefield.

León-Portilla, M. (1996). *El destino de la palabra. De la oralidad y los códices mesoamericanos a la escritura alfabética* [Destiny of the world: From orality and meso-American codices to the written alphabet]. México City, Mexico: El Colegio Nacional Fondo de Cultura Economica.

Marcus, J. (1992). *Mesoamerican writing systems: Propaganda, myth, and history in four ancient civilizations.* Princeton, NJ: Princeton University Press.

Munir, K. (2002). Being different: How normative and cognitive aspects of institutional environments influence technology transfer. *Human Relations, 55,* 1403–1428.

Orlikowski, W. (1992). The duality of technology: Rethinking the concept of technology in organizations. *Organization Science, 3,* 398–427.

Osland, J., de Franco, S. & Osland, A. (1999). Organizational implications of Latin American culture: Lessons for the expatriate manager. *Journal of Management Inquiry, 8,* 219–237.

Rosenn, K. (1988). A comparison of Latin American and North American legal traditions. In L. Tavis (Ed.), *Multinational managers and host government interactions* (pp. 127–152). South Bend, IN: University of Notre Dame Press.

Stewart, E., & Bennett, M. (1991). *American cultural patterns: A cross-cultural perspective* (Rev. ed.). Yarmouth, ME: Intercultural Press.

Sullivan, P., & Porter, J. (1997). *Opening spaces: Writing technologies and critical research practices.* Greenwich, CT: Ablex.

Thatcher, B. (1999). Cultural and rhetorical adaptations for South American audiences. *Technical Communication, 46,* 177–195.

Thatcher, B. (2000). L2 professional writing in a U.S. and South American context. *Journal of Second Language Writing, 9*(1), 41–69.

Thatcher, B. (2001). Issues of validity in intercultural professional communication research. *Journal of Business and Technical Communication, 15,* 458–489.

Thatcher, B. (2005). Situating L2 writing in global communication technologies. *Computers and Composition, 22,* 279–295.

Trompenaars, F., & Hamden-Turner, C. (1998). *Riding the waves of culture: Understanding diversity in global business* (2nd ed.). New York: McGraw Hill.

Victor, D. (1992). *International business communication.* New York: HarperCollins.

Yin, R. (1994). *Case study research: Design and methods* (Rev. ed.). Newbury Park, CA: Sage.
Zuboff, S. (1989). *In the age of the smart machine: The future of work & power.* New York: Basic
Books.

Barry Thatcher is associate professor of rhetoric and professional communication at
New Mexico State University. His research interests include intercultural professional
communication, history of rhetoric in Mexico and Latin America, and intercultural research methods.

TECHNICAL COMMUNICATION QUARTERLY, *15*(3), 407–410

REVIEW

Tracy Bridgeford, University of Nebraska at Omaha, Editor

Rhetoric, Innovation, Technology: Case Studies of Technical Communication in Technology Transfers. **Stephen Doheny-Farina. Cambridge, MA: MIT Press, 1992. 279 pp.**

Reviewed by Ann Brady
Michigan Technological University

I sit at my desk, preparing for a new class that introduces undergraduate technical communication students to a complex communication process that many will find themselves involved in professionally: technology transfers. Although it has been 15 years since *Rhetoric, Innovation, Technology: Case Studies of Technical Communication in Technology Transfers* was first published, I find myself returning to it—and ultimately adopting it for my class because what its author, Stephen Doheny-Farina, demonstrated 15 years ago is as useful now as it was then. Technology transfers, which take laboratory research results through the production and marketing process and then adapt them to place them in the hands of users, are not decontextualized transfers of facts. Instead, they are a web of complex rhetorical negotiations among designers, developers, writers, and users who support the construction of new knowledge and information. Some technical writers at a few companies may find themselves included in the early stages of a product cycle. Many writers, however, are still marginalized, perceived as serving technical experts and thus easily left out of production cycles until the product is ready for market when they are then expected to take what's given them and write it up. Technical communicators must continue to make the argument that because they mediate between the worlds of technology and those of end users, their involvement in all stages of product design, development, and documentation is crucial. Doheny-Farina's book gives these fledgling professionals the awareness—and some of the evidence—they need to wage those arguments more successfully.

The book is organized into five chapters. Chapter 1 offers three scenarios that demonstrate the complex and dynamic social processes that move information from laboratories to the marketplace. Pointing out the rhetorical nature of these processes, Doheny-Farina critiques conventional notions of communication as a one-way movement from expert to user that focuses on the facts being transferred and omits entirely the difficulties that people have in understanding those facts. Proposing a more finely textured view of communication—one that advances the argument that all participants in information exchange contribute to the construction of new knowledge about new technologies—Doheny-Farina's next four chapters build on the first. Not only do they report detailed case studies that illustrate the rhetorical nature of technology transfers, but they are also engaging reading because they show us how theories of rhetoric and communication play out in the often turbulent sites of the workplace. Chapter 2, for instance, demonstrates how the future of an entrepreneurial start-up company is contested—and ultimately constructed—not by high-level discussions in boardrooms with venture capitalists, but by a small group of individuals negotiating, interpreting, and adjusting their understandings of their company's business plans: what they hope will become a common vision of the company's mission.

Both in location and in intent, chapter 3 offers a different view of technology transfer. Here, the focus is on university and industry collaboration, thus demonstrating the difficulties of spanning boundaries across organizations—by contrast to within organizations—as information about a groundbreaking, artificial heart is moved from technical developers and designers to writers and finally to users. Building on the complexities of technology transfers across organizations, chapter 4 presents two cases, both of which call for rhetorically trained technical communicators to be involved in the early stages of product development, to negotiate the shifting boundaries that separate designers from marketers, marketers from users, and thus adapt new technologies to users earlier and more successfully. The last chapter discusses concepts crucial in preparing technical communication students to go beyond conventional notions of their future work as documenting what others have created. Here, Doheny-Farina argues that preparing students to communicate with future colleagues and coworkers who have been educated in different disciplines means exposing them to theories of rhetoric, enculturation, and discursive practice. Only with such education will upcoming technical communicators learn how to work as mediators within work groups and across complex and often conflicting organizational alliances. Supplementing the last chapter in particular, the book's appendix includes a practical discussion of specific applications for the technical communication curriculum and a detailed scenario intended for classroom use, foregrounding the challenges of collaboration within organizations.

Quite clearly, *Rhetoric, Innovation, Technology* is a rich source for classroom discussions about a range of topics. It is true that the book focuses exclusively on how texts shape organizational identities and the construction of knowledge within organizational walls—and that this is a drawback. However, there are ways to use the book to encourage students to see how organizations themselves are shaped by larger economic and political systems beyond their boundaries.

Gender, race, and class, for instance, do not figure in the intra-organizational communication conflicts that the designers, developers, and technical writers experience in the book's case studies and scenarios. In our field, some progress has been made in examining the gendered relations of technology (see Lay, 1991, for instance). However, the absence of race identity and race markers continues to normalize the nature of communication in and across organizational groups and corporate entities. Unfortunately, *Rhetoric, Innovation, Technology* does little to call attention to this fact. However, this does not keep classroom teachers from invoking the innovative spirit of the book's case studies and scenarios. For example, teachers might lead a discussion that requires students to think of Liz Cates, noted in the case of the entrepreneurial start-up company detailed in chapter 2, not as an unmarked, generic, white woman, but as an African-American. Doing so could lead to discussions about how race and gender intersect in this case to complicate the struggle over who speaks and who does not in making decisions about where the company is headed. Such discussions would surely enrich technical communication students' understanding of the complexities of information transfer. Technical communication students could also benefit from more discussions about organizational classes than the book offers. Because students may very well find themselves on the lower rungs of organizational hierarchies, classed as scribes who make documents look pretty and read well, they must be prepared to negotiate their positions, to explain why they should be recognized as fully qualified participants in the process of product development and information transfer. Although Doheny-Farina's book is an extended argument for this position, his case studies present such positive collaborations among writers, engineers, and scientists that important questions about organizational class are never raised: Who is the "authority" in these relations? Can a technical communicator, in fact, assume a position of authority (see Slack, 2003)? How might authority be contested or negotiated; how does organizational class play into such negotiations? Fortunately, again, the book offers moments when a classroom teacher can raise such questions and foreground them for discussion. So, for instance, in chapter 4, one technical writer reports that to gain the respect of her technical colleagues on a design team, she had worked no less than 12 hours/day and often as many as 16 or 18, writing and looking at code:

> I felt it was the first time that Technical Development had looked at us as an equal partner in the development process. It was the first time that they looked at us as anything other than a secretary who just simply took what they had done and rewritten it to suit our needs. We were finally thinking people. (p. 194)

Doheny-Farina comments that technical writers like these pay a price for joining a team like this early and establishing their authority on it, but adds, "It took that amount of effort to gain the respect of the developers and become a substantive contributor to the team" (p. 194). Here, I would want to ask my students the following questions: What kind of knowledge authorizes technical communicators in this case? Are there other types of knowledge that technical communicators have to offer? What does it mean to be a thinking person in an organization?

Although I have decided to use *Rhetoric, Innovation, Technology* in my undergraduate class, I am also considering it for a graduate class in technical communication and technology studies. Here, the book would raise different but intersecting questions from those posed for undergraduates. Let me explain. Working from a distinction between technical communication as a skill or art—a *techne*—and technical communication as a practice intended for social action—a *praxis*—Doheny-Farina makes the point that "*techne* implies work to produce a well-crafted document, a necessary and important ability. *Praxis* implies work to produce a social good" (pp. 219–220). Although the author takes up questions of who constitutes the social group, what is the social good, and what role does the technical communicator play in advancing it, he also suggests that only praxis can move us toward significant social ends. In the past 15 years, scholars such as Johnson (1998) have marked out the intersections of *techne* and *praxis,* suggesting that the production of a document, if it is designed with the user in mind, works to produce a social good as well. My questions to graduate students, arising from Doheny-Farina's text, would ask about the epistemological nature of *techne* and *praxis,* how, and if, the two differ, and how they might intersect to advance responsible social action.

Since Doheny-Farina's book was published 15 years ago, our field has seen many changes, among them, thankfully, a more focused interest in issues of gender, race, class, and what it means to practice technical communication in a socially responsible way. I hope for a *Volume Two of Rhetoric, Innovation, Technology,* which would surely pick up these issues where the first left off.

In the meantime, introducing Doheny-Farina's work to undergraduate students will both prepare them for the challenges they face as socially responsible professionals and give them insights into how they can locate themselves assertively in complex communication situations. *Rhetoric, Innovation, Technology* also offers valuable insights into how rhetoric operates tacitly in the workplace for the more advanced students or for those interested in examining theories of how knowledge is constructed there.

REFERENCES

Johnson, R. R. (1998).*User-centered technology: A rhetorical theory for computers and other mundane artifacts.* Albany: State University of New York Press.

Lay, M. M. (1991). Feminist theory and the redefinition of technical communication. *Journal of Business and Technical Communication, 5,* 348–370.

Slack, J. D. (2003). The technical communicator as author: A critical postcript. In T. Kynell-Hunt & G. Savage (Eds.), *Power and legitimacy in technical communication: Vol. I. The historical and contemporary struggle for professional status* (pp. 193–207). Amityville, NY: Baywood.